Surfactant-based Sensors in Chemical and Biochemical Detection

Detection Science Series

Editor-in-chief
Mike Thompson, *University of Toronto, Canada*

Series editors:
Subrayal M. Reddy, *University of Surrey, UK*
Damien Arrigan, *Curtin University, Perth, Australia*
Mengsu (Michael) Yang, *City University of Hong Kong, Hong Kong*

Editorial advisor:
Electra Gizeli, *IMBB-FORTH, Greece*
Thiago Paixão, *University of São Paulo, Brazil*

Titles in the Series:
1: Sensor Technology in Neuroscience
2: Detection Challenges in Clinical Diagnostics
3: Advanced Synthetic Materials in Detection Science
4: Principles and Practice of Analytical Techniques in Geosciences
5: Microfluidics in Detection Science: Lab-on-a-chip Technologies
6: Electrochemical Strategies in Detection Science
7: Peroxynitrite Detection in Biological Media: Challenges and Advances
8: Biological Fluid-Surface Interactions in Detection and Medical Devices
9: Advanced Environmental Analysis: Applications of Nanomaterials, Volume 1
10: Advanced Environmental Analysis: Applications of Nanomaterials, Volume 2
11: Quenched-phosphorescence Detection of Molecular Oxygen: Applications in Life Sciences
12: Carbon-based Nanomaterials in Analytical Chemistry
13: Forensic Analytical Methods
14: Immunosensors
15: Analytical Electrogenerated Chemiluminescence: From Fundamentals to Bioassays
16: *In Situ* Analysis of Cellular Functional Molecules
17: Confining Electrochemistry to Nanopores
18: Detection Methods in Precision Medicine
19: Analytical Strategies for Cultural Heritage Materials and their Degradation

20: Challenges in Detection Approaches for Forensic Science
21: Disposable Electrochemical Sensors for Healthcare Monitoring
22: Quantum Dots in Bioanalytical Chemistry and Medicine
23: Surfactant-based Sensors in Chemical and Biochemical Detection

How to obtain future titles on publication:
A standing order plan is available for this series. A standing order will bring delivery of each new volume immediately on publication.

For further information please contact:
Book Sales Department, Royal Society of Chemistry, Thomas Graham House, Science Park, Milton Road, Cambridge, CB4 0WF, UK
Telephone: +44 (0)1223 420066, Fax: +44 (0)1223 420247
Email: booksales@rsc.org
Visit our website at www.rsc.org/books

Surfactant-based Sensors in Chemical and Biochemical Detection

Edited by

J. G. Manjunatha
FMKMC College, Constituent College of Mangalore University, India
Email: manju1853@gmail.com

ROYAL SOCIETY OF CHEMISTRY

Detection Science Series No. 23

Print ISBN: 978-1-83767-116-8
PDF ISBN: 978-1-83767-118-2
EPUB ISBN: 978-1-83767-119-9
Print ISSN: 2052-3068
Electronic ISSN: 2052-3076

A catalogue record for this book is available from the British Library

The Royal Society of Chemistry is a charity, registered in England and Wales, Number 207890, and a company incorporated in England by Royal Charter (Registered No. RC000524), registered office: Burlington House, Piccadilly, London W1J 0BA, UK, Telephone: +44 (0)20 7437 8656.

For further information see our website at www.rsc.org

Printed in the United Kingdom by CPI Group (UK) Ltd, Croydon, CR0 4YY, UK

Preface

This book describes surfactant-based sensors in electrochemistry and the importance of surfactants in electrochemical investigations. The chapters, contributed by various researchers, furnished the progression of highly responsive electrochemical sensors utilizing surfactants as a modifying agent to the sensor field and information concerning the activity of target molecules and electron transfer kinetics arising at the fabricated sensing surface.

The book presents the applications of surfactants for electrochemical sensors. The focus of this book is to attract the scientific community, including students, researchers, academics (including professors from the different fields of electrochemistry), industrial scientists, engineers, and medics, and to stimulate more researchers to construct highly sensitive devices based on high-response surfactant and electrode materials. This book will also help to improve competence among people and researchers from different fields, such as physics, electronics, chemistry, engineering, and biology, who are expected to collaborate on innovative research in this newly emerging interdisciplinary research area, *i.e.* surfactant-based electrochemical sensors in various applications.

J. G. Manjunatha

Detection Science Series No. 23
Surfactant-based Sensors in Chemical and Biochemical Detection
Edited by J. G. Manjunatha

Published by the Royal Society of Chemistry, www.rsc.org

Acknowledgements

I gratefully acknowledge my sincere gratitude to my mother, Sharadamma Jamballi, wife, Sinchana Manjunath, daughters, Nirvikaa J. M. and Vruddhi J. M., and research colleagues.

Detection Science Series No. 23
Surfactant-based Sensors in Chemical and Biochemical Detection
Edited by J. G. Manjunatha

Published by the Royal Society of Chemistry, www.rsc.org

Contents

Detection Science Series No. 23
Surfactant-based Sensors in Chemical and Biochemical Detection
Edited by J. G. Manjunatha

Published by the Royal Society of Chemistry, www.rsc.org

CHAPTER 1

Review of Surfactants, Structural Properties and Their Role in Electrochemistry

DIDEM NUR UNAL,[a,b] SELENAY SADAK,[a,b] CEM ERKMEN,[a] ÖZGE SELCUK[a,c] AND BENGI USLU*[a]

[a] Department of Analytical Chemistry, Ankara University, Faculty of Pharmacy, 06560, Ankara, Turkey; [b] The Graduate School of Health Sciences, Ankara University, 06110, Ankara, Turkey; [c] Department of Analytical Chemistry, Faculty of Pharmacy, Mersin University, 33169, Mersin, Turkey
*Email: Buslu@pharmacy.ankara.edu.tr

1.1 Introduction

Surfactants, also called surface active agents, which have an amphiphilic structure, consist of two parts, the hydrophilic head and hydrophobic tail.[1]

These organic compounds, due to their unique structure, are soluble in both organic solvents and water. These properties come from their ability to form micelles in solution. Their ability to form micelles also makes it possible to use them as detergents. In addition, this situation causes surfactants to accumulate at the interface of two insoluble liquids, reducing the interfacial tension between the liquids and increasing the solubility of the two liquids in each other. Surfactants exist as monomers when dissolved in water at low concentrations. When the concentration increases and the optimum concentration is reached, the monomers come together to form micelles. When monomers come together, their hydrophobic tails are located towards the center of the micelle where hydrophilic heads settle

Detection Science Series No. 23
Surfactant-based Sensors in Chemical and Biochemical Detection
Edited by J. G. Manjunatha

Published by the Royal Society of Chemistry, www.rsc.org

towards the solvent. With the formation of micelles, the free energy and surface tension of the system decrease and this decrease is fixed at one point. The concentration of surfactant monomer required for the formation of micelles is called the critical micelle concentration (CMC). The degree of micellization of surfactants depends on the chain length of the hydrophobic group in their structure. This value decreases as the hydrophobicity increases.[2,3]

Surfactants have applications in a wide variety of areas of chemistry. These amphiphilic molecules have been used in nanotechnology,[4,5] pharmaceutical formulations,[6–8] detergents,[9,10] corrosion inhibitors,[11–13] emulsifier systems,[14,15] and petroleum industry.[16]

In addition to this wide area of use, surfactants have had a wide variety of applications in the field of electrochemistry from past to present.[17] Due to the adsorption and aggregation properties of surfactants at the solution interface, they have found use in areas such as catalysis, capacitors, corrosion and electrochemical sensors.[18–20] For example, lowering the interfacial tension of surfactants can result in better wettability for the electrodes, and improved charge accumulation can be achieved by using surfactants as the electrolyte in capacitors.[21] In corrosion studies, corrosion resistance can be increased by changing the surface morphology of the coating by adding surfactants to the coating of surfaces with nanomaterials[22] and they can be used as corrosion inhibitors.[23–25] They are frequently used in electrochemical sensors to modify the working electrode or to increase the electron transfer rate in the electrolyte solution, to get more sensitive responses by expanding the surface area, or to increase the response of the analyte by using them with different nanomaterials.[17,26–30]

1.2 Structures and Features of Surfactants

Surfactants are chemical molecules that are active at the surfaces or solution interface between two immiscible phases. A hydrophilic head and a hydrophobic tail in the structure of this molecule give them unique properties (Figure 1.1). While the hydrophobic tail is apolar and dislikes water, the hydrophilic polar group likes it. These properties enable surfactants to be amphiphilic compounds soluble in both water and organic solvents.[31,32] The hydrophobic tails are usually long hydrocarbon chains of alkyl or acyl benzene-type methyl, methylene or phenol groups.[2,9,13]

These amphiphilic molecules are oriented to the surfaces and interfaces between the two phases according to the solution properties. This helps to reduce the surface tension between the two phases.[32] Surfactants begin to aggregate into clusters called micelles when saturated at sufficiently high concentration between the two phases, known as the CMC[34] and which can be determined by different techniques.[35,36] The CMC values at which micelle structures are formed are a very important value in the use of surfactants, because the molecules in the environment can perform different interactions with aggregated surfactants.[37] Aggregation number (N_{agg}) is a term

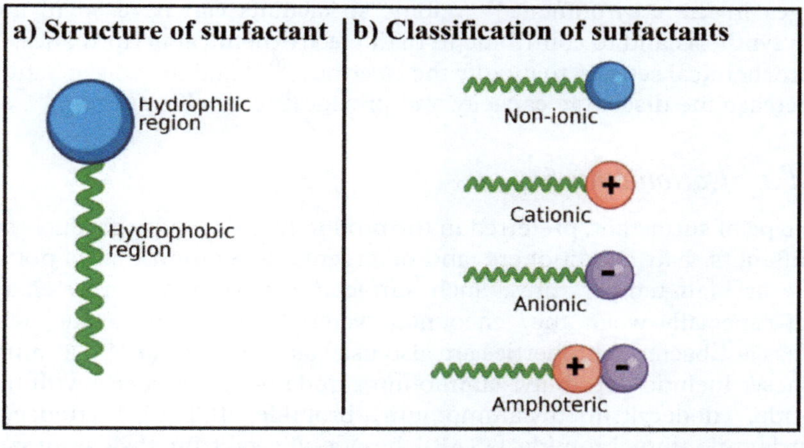

Figure 1.1 Structure and classification of surfactants. Reproduced from ref. 33, https://doi.org/10.3390/ma14123197, under the terms of CC BY 4.0 license, https://creativecommons.org/licenses/by/4.0/.

used to describe the total number of micelles. The N_{agg} value depends on the type and structure of the surfactant, the electrolytes added, the nature of the solvent, the pH of the medium, and the temperature. In addition, micelles can appear in different ways, depending on these conditions.[32]

Although the use of a single surfactant is sufficient to reduce micelle formation and surface tension, it may be necessary to use surfactant blends to further reduce surface tension and the CMC. Above the CMC level, surfactants cause hydrophobic organic compounds to dissolve in higher concentrations than when they are alone in water.

1.2.1 Classification of Surfactants

To understand the use of surfactants in various applications, their classification is necessary. Although they serve a common purpose, surfactants differ in structure and are divided into four groups according to the charge carried by the hydrophilic head part of their molecular structure. These are anionic surfactants, cationic surfactants, non-ionic surfactants and amphoteric surfactants[9] (Figure 1.1).

1.2.1.1 Anionic Surfactants

These are chemically stable and highly soluble, and are widely used in the chemical industry. Examples of these most common surfactant groups are sodium dodecyl sulfate (SDS), sodium lauryl sulfate (SLS) and gemini. Surfactants in this group consist of a hydrophobic part, an alkyl chain of varying length, and a hydrophilic part that acquires a negative charge in aqueous solutions. Because they are negatively charged, they interact with positive charges in the environment.[38] Anionic surfactants can be used in metal oxide synthesis and to contribute to their electrochemical performance,[39] in electrochemical sensors to modify the electrode,[40,41] and in lithium batteries to increase the discharge capacity and provide energy efficiency.[42,43]

1.2.1.2 Cationic Surfactants

This type of surfactant, preferred in the production of laundry products such as softeners, hair conditioners, and detergents, is a group that is not very stable at high temperatures. Such surfactants have a positively charged head, especially when they encounter water. Cationic surfactants which possess antibacterial properties are also used as disinfectants.[38,44] Examples of these include tetrabutyl ammonium rodanide, *N*-dodecylpyridinium chloride, dodecyltrimethylammonium bromide (DTAB), tetradecyltrimethylammonium bromide (TTAB), hexadecyl(cetyl)trimethyl ammonium bromide (CTAB), *etc.*[45] Cationic surfactants such as CTAB have been used in the electrochemical CO_2 reduction reaction, promoting carbon product formation and have been reported to suppress hydrogen formation.[46,47] The interfacial field produced by surfactants affects the energies of the electrochemical intermediates, necessitating their detailed understanding.

1.2.1.3 Non-ionic Surfactants

These have been the subject of many studies and are a group of affordable and extremely tolerant surfactants that act neutrally in aqueous solutions. Since they do not have an electronic charge, they dissolve in water by forming hydrogen bonds. These properties cause a decrease in their stability at high temperatures. Examples are Triton X-100, polyvinylpyrrolidone, and saponin. Triton X-100 can be used as an electrolyte in plating to achieve high specific capacitance by electroprecipitation,[48] in electrocrystallization,[49] in electrochemical sensors[50] and as a corrosion inhibitor.[51]

1.2.1.4 Amphoteric or Zwitterionic Surfactants

The last group of surfactants have both positive and negative charges. Their low toxicity, and eye and skin compatibility have led them to be used in many cosmetic and pharmaceutical products. These surfactants are of the hydroxypropyl carboxybetaine or sulfobetaine type, or amido sulfobetaine or amido caboxybetaine type if a hydroxyl group is present on the segments between the two charged groups.[52,53] Amphoteric surfactants include sulfobetaine, carboxybetaine, sulfoimidazolium, and carboximidazolium.[53] The presence of both anionic and cationic groups has enabled properties such as increased water solubility,[54] low toxicity,[55] easy biodegradation,[56] and stability against temperature changes.[53,57] Due to their structure, amphoteric surfactants can be used as additives to the electrolyte solution in batteries, to inhibit corrosion,[51,58,59] as stabilizing agents in battery systems[58,60] and to modify the working electrode in electrochemical sensors and biosensors.[61–63]

Surfactants can also be classified according to the source from which they are produced, such as synthetic and bio-surfactants. The majority of synthetic surfactants produced through chemical reactions are copolymers of ethylene oxide and propylene oxide, alkoxylated alkylphenol derivatives, sorbitan esters, alcohols, and amines. Bio-surfactants are based on biological material and are considered less toxic and environmentally friendly than synthetic varieties.[64,65] Synthetic and bio-surfactants have various chemical properties associated with their structural differences and have been the subject of many different application areas and electrochemical studies based on these properties.[16,17,51,66–69]

1.3 Applications of Surfactants in Electrochemistry

1.3.1 Electroplating

In electrochemistry studies, the first use of surfactants is usually in electroplating processes. Surfactants are often used to clean metal surfaces or accelerate hydrogen generation at the cathode. Fatty alcohol exoxylates, sulfates and alkyl benzene sulfonates are the most preferred surfactant groups for this purpose. Surfactants can provide many advantages in electroplating. By using surfactants, the operating range can be extended with less influence from environmental conditions such as pH, temperature, and current density. Separation of gas bubbles from the cathode and cleaning of possible deposits from metal surfaces becomes easier. In this way, the use of surfactants allows the reduction of surface tension.[17,70]

In a study by Lv *et al.*, various fatty acid quaternary ammonium surfactants derived from renewable fatty acid sources were designed for use as applied stabilizers in copper electroplating. The surface tension and contact angle experimental results showed that the surface activity of surfactants is closely related to the length of the hydrophobic chains. The ability of the developed surfactants to prevent copper deposition was investigated by electrochemical methods. Among the synthesized compounds, it was extensively demonstrated by field emission scanning electron microscopy (FE-SEM), X-ray photoelectron spectroscopy (XPS) and other electrochemical measurements that the compound with an $R = C_{15}H_{31}$ functional group performed best in the inhibition of copper. The inhibition ability of this compound is due to its fast and strong absorption onto the cathode surface. In addition, the galvanostatic measurement results showed that this compound has stable synergistic effects with other additives to form a competitive adsorption layer due to convection.[71]

In a study by Shih *et al.*, the crystal structure of metallic tin (Sn) in a nickel foam electrode (Sn/Ni) and the selective conversion of NO^{3-} were evaluated in the presence of cationic surfactants. The Sn electroplating additive contained different surfactants, such as benzetonium chloride (BZT), cetylpyridinium chloride (CPC), cetyltrimethylammonium chloride (CTAC) and hexyltrimethylammonium bromide (C_6TAB). The characterization results showed that the efficiency of nitrate reduction on the Sn surface was high in the presence of BZT.[72]

1.3.2 Corrosion

Corrosion is described as the degradation brought on by contact of a material (usually a metal) with various environmental conditions. It is a common potential risk for oil and gas production, or transportation facilities. Although the corrosion reaction cannot be stopped, slowing it down can lead to a cost-effective solution. Surfactants have been extensively studied as corrosion inhibitors in recent years.[73] Surfactants are preferred as corrosion inhibitors over other metal inhibitors through their amphiphilic nature, cheapness, low toxicity, and resistance to metal corrosion by adsorption.[51]

In their study, Bedair *et al.* prepared three different non-ionic surfactants based on azodi and Schiff bases. They investigated their abilities as corrosion inhibitors for steel in 1.0 M HCl by using electrochemical frequency modulation, potentiodynamic polarization and electrochemical impedance spectroscopy methods. As a result, they concluded that the adsorption of the surfactants they developed on the carbon steel surface in 1.0 M HCl was compatible with the Langmuir isotherm.[74]

In another study, Shaban *et al.* synthesized non-ionic dithiol surfactants using ethylene glycol containing sulfur atoms. They investigated the potential of these surfactants as corrosion inhibitors for mild steel in 1 M HCl solution. In order to evaluate the inhibition effects of surfactants, weight loss at three different temperatures was investigated using potentiodynamic and impedance methods. As a result, the adsorption process of the synthesized surfactants on the metal steel surface was found to be compatible with the Villamil adsorption isotherm.[75]

In 2023, Abd-ElHamid *et al.* synthesized cationic surfactants with azomethine groups that could be used as both corrosion inhibitors and biocides. This study aimed to provide maximum effectiveness on steel or bacterial surfaces by integrating azomethine groups into the gemini surfactant structure. While the properties of the synthesized structures were investigated by Fourier-transformed infrared spectroscopy, nuclear magnetic resonance spectroscopy and mass spectroscopy, the effects of corrosion on mild steel in 1.0 M HCl were investigated gravimetrically and electrochemically. As a result, the synthesized gemini cationic surfactant-based structure was shown to act as a bactericide against sulfate-reducing bacteria, which are one group of corrosion-causing bacteria.[76]

In another study by Jalab *et al.*, the C-steel corrosion prevention performance of Quaternium-22 (Q-22), an environmentally friendly surfactant, which is frequently used in the cosmetics industry, was investigated in 5 M HCl. It was found that 2.22 mmol L^{-1} Q-22 exhibited maximum efficiency when weight loss, concentration, and immersion times were optimized. While the anticorrosion effect of Q-22 on C steel was elucidated by electrochemical methods, the adsorption results obtained, which followed the Freundlich isotherm, and the physisorption results of Q-22 on the C-steel surface were found to be in agreement. In addition, the relative inhibition performance of Q-22 on the C-steel surface was also investigated using scanning electron microscope (SEM) and atomic force microscope (AFM) images. The surface morphology seen in the SEM images and changes in the presence and absence of Q-22 inhibitor confirmed the inhibitory effect of Q-22 (Figure 1.2(A)). In addition, the roughness of the metal surface (RMS) value, which changed when Q-22 inhibitor was used on the steel surface, also confirmed the effect of the Q-22 inhibitor (Figure 1.2(B)).[77]

1.3.3 Batteries

Recently, the use of environmentally friendly and highly efficient fuel cells has been in great demand to replace rechargeable batteries in electronic devices used for various purposes, including medical and military. Nanoparticle-based catalysts containing surfactant significantly increase the performance of the oxidation or reduction reaction and the catalytic activity of the developed catalysts.[78]

In their work, Yang *et al.* synthesized PdBi-based bimetallic catalysts. They evaluated the catalytic activities of these catalysts, the stability of the formic acid oxidation reaction, and their performance directly in a formic acid fuel cell. While synthesizing this catalyst, the researchers used amphiphilic polyvinylpyrrolidone surfactant to reduce aggregation. The catalytic activity and morphology of the catalyst containing the optimal amount of surfactant was investigated optically and electrochemically using transmission electron microscopy (TEM), XPS, cyclic voltammetry (CV), and polarization methods. The obtained results showed that the developed catalyst exhibited long-term stability and excellent catalytic activity.[79]

In the work of Xing *et al.*, nitrogen and graphene (NG)-doped Co_3O_4 nanoparticles were produced for the first time by a surfactant-assisted hydrothermal method. Three different surfactants, including triblock copolymer F127, cetyltrimethyl ammonium bromide, and sodium dodecyl sulfate, were used in the synthesis of these nanoparticles. Under optimal conditions, triblock copolymer F127-supported NG/Co_3O_4 composite as anode material in a lithium ion battery showed superior performance compared to most other anode materials.[43]

In another study, a uniform $LaSrCoO_4$ material was prepared using an amino-acid-derived surfactant (*N*-(2-hydroxylacyl)-1-phenylalanine). The rich active groups of *N*-(2-hydroxylacyl)-1-phenylalanine allowed the developed $LaSrCoO_4$-based material to exhibit improved specific surface area and enhanced electrocatalytic properties. The results obtained under optimum conditions confirmed that $LaSrCoO_4$ can be used as an ideal cathode material for solid oxide fuel cells.[80]

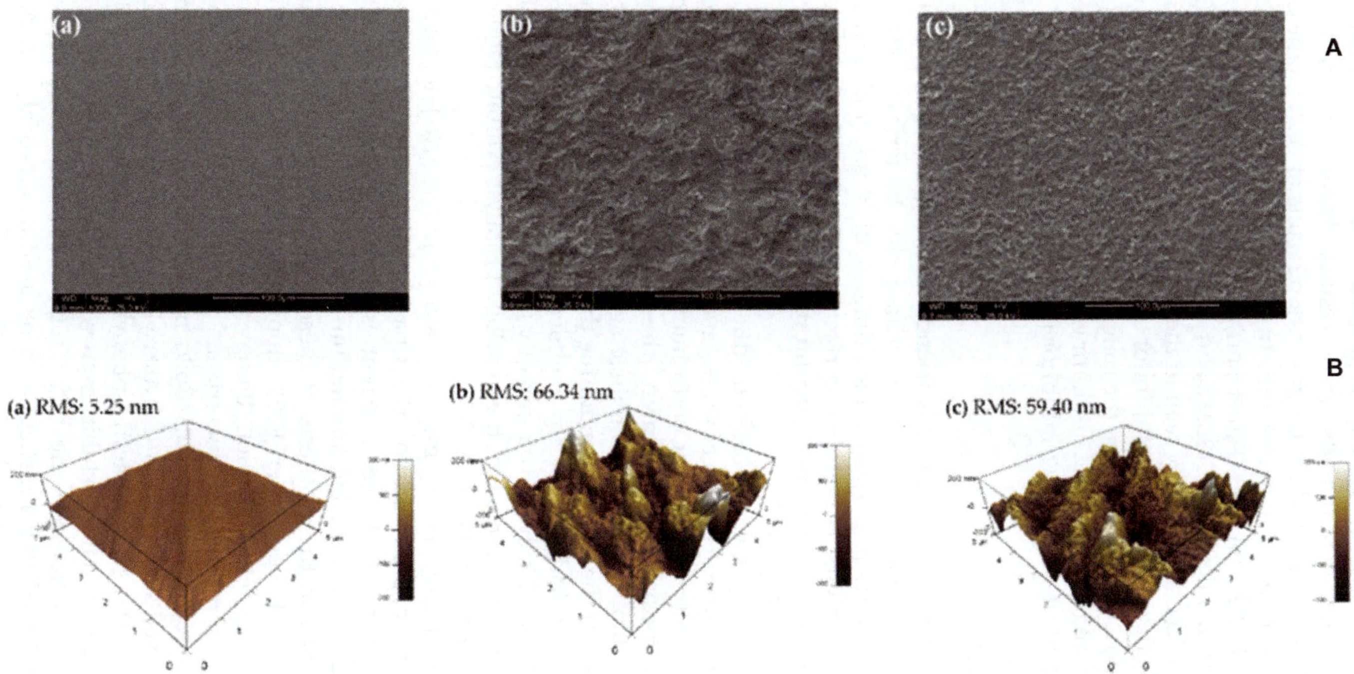

Figure 1.2 (A) SEM micrographs of (a) a polished C-steel coupon in the (b) absence and (c) presence of 2.22 mmol L^{-1} Q-22 inhibitor. (B) AFM images of (a) a polished C-steel coupon in the (b) absence and (c) presence of 2.22 mmol L^{-1} Q-22 inhibitor. Reproduced from ref. 77, https://doi.org/10.3390/molecules27196414, under the terms of CC BY 4.0 license, https://creativecommons.org/licenses/by/4.0/.

There is a need to develop synthesis strategies to improve the general properties of polymeric materials such as polyaniline and polypyrrole for use in energy storage applications. Among polymeric materials, polybenzopyrrole (Pbp) stands out as a good candidate for use in electrochemical energy conversion and storage. In their study, Begum *et al.* synthesized Pbp in an acidic medium supplemented with surfactant. The physicochemical and electrochemical properties of these synthesized materials were investigated in order to be used as electrode material. The anionic-surfactant-assisted polymerization reaction for the synthesis of Pbp in this study is shown in Figure 1.3. After optimizing various reaction parameters, the synthesized material showed thermal stability up to 549.9 °C, while significant improvement was achieved in its electrochemical properties. Under optimal conditions, the synthesized material exhibited a current density of 1 A g^{-1} and an excellent cycling stability of 87% after 5000 cycles. In addition, this material had a maximum specific capacitance of 166.0 ± 2.0 F g^{-1}, and a high power density of 2.75 kW kg^{-1} was obtained for the surfactant-assisted synthesized polybenzopyrrole. These results showed that the surfactant-assisted synthesized materials can be used as electrode materials in next-generation Faradaic supercapacitors.[81]

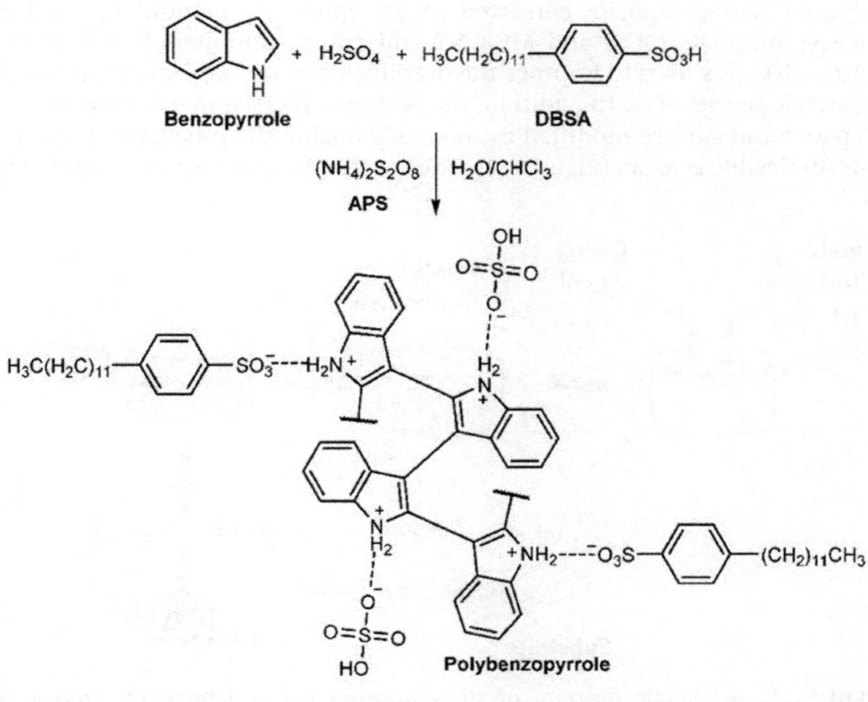

Figure 1.3 Schematic synthesis of Pbp in H_2SO_4/DBSA. Reproduced from ref. 81, https://doi.org/10.3390/polym13172883, under the terms of CC BY 4.0 license, https://creativecommons.org/licenses/by/4.0/.

Surfactants can also be used as wetting agents during several electrometallurgical processes, such as electrowinning and electroplating, in order to increase current efficiency and deposit quality. In the Shu *et al.* and Tian *et al.* studies, they suggested the addition of surfactant additives to the medium for the extraction of electrolytic manganese residue, which is the acid leach residue of manganese carbonate ores containing manganese and ammonia nitrogen, which causes serious damage to the environment.[82–84]

1.3.4 Electrochemical Sensors

Electrochemical sensors are a another application for surfactants that has gained attention recently. The use of modified electrodes in electrochemical sensors with advantages such as increasing electron transfer rate, amplification of electrochemical signals, and effects on redox potentials has drawn attention. In addition, the presence of anionic and cationic groups in their structures makes them suitable molecules to be included in the electrolyte solution composition in electrochemical sensors.[27,63]

In their work, Nag *et al.* performed the preparation of multi-walled carbon nanotubes (MWCNTs)/polydimethylsiloxane (PDMS) nanocomposites in the presence of surfactant and their application in tactile determination. The prepared nanocomposite consisted of an optimized amount of sodium dodecyl sulfonate (SDS) and MWCNTs mixed as nanofillers in the PDMS matrix. SDS was used to improve the overall electromechanical properties of the developed sensors. In addition, the prepared nanocomposite was simply prepared and surface modified by manually mixing the processed materials to form flexible sensors (Figure 1.4). The synergistic effect between MWCNTs

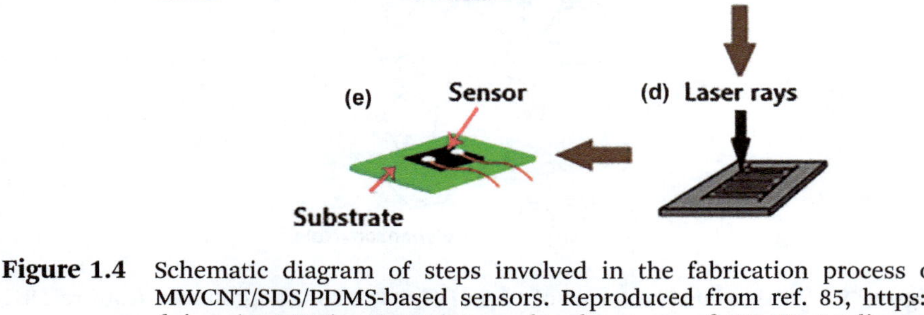

Figure 1.4 Schematic diagram of steps involved in the fabrication process of MWCNT/SDS/PDMS-based sensors. Reproduced from ref. 85, https://doi.org/10.3390/ma15134504, under the terms of CC BY 4.0 license, https://creativecommons.org/licenses/by/4.0/.

and SDS which acted as a surfactant, resulted in a decrease in the resistance value. In addition, more sensitive results were obtained with the increase in electrical conductivity.[85]

In their study, Sakač *et al.* synthesized a new 1,3-dihexadecyl-1*H*-benzo [*d*]imidazole-3-ium–tetraphenylborate (DHBI–TPB) surfactant and used it as a modification agent for the determination of cationic surfactants in personal care products and disinfectants. The synthesized material was successfully characterized using nuclear magnetic resonance (NMR), Fourier transform infrared (FTIR) spectrometry, liquid chromatography–mass spectrometry (LC–MS) and elemental analysis methods. The sensor, whose electrochemical properties were improved using DHBI–TPB, was successfully applied for the determination of CTAB, cetylpyridinium chloride (CPC) and hyamine from 12 different commercial personal care products and disinfectant samples.[86]

In another research study, a voltammetric method was developed for the determination of trimethoprim (TMP) using a nanostructured zinc oxide nanoparticle-modified carbon paste electrode (ZnO/CPE) containing SDS. While the developed electrode containing SDS exhibited the highest peak current at 1.0 mM $K_3(Fe(CN)_6)$ (Figure 1.5(A)) compared to bare CPE and ZnO/CPE, the highest peak current was obtained for 0.5 mM TMP in 2 M PBS with the developed electrode (Figure 1.5(B)). With increased conductivity and facilitation of electron transfer, linearity was achieved for TMP in the concentration range 8.0×10^{-7}–1.0×10^{-5} M. The lower limit of detection was found to be 2.58×10^{-8} M. As a result, the developed sensor was able to show superior performance with excellent recovery values for the determination of TMP in pharmaceutical dosage forms and urine samples.[26]

1.4 Conclusion and Future Prospects

Surfactants include two separate components in their structures, polar and apolar groups, in the same structure, and are surface-active agents with amphiphilic qualities that reduce the surface tension of a liquid. They are often of organic origin. Each system has certain distinctive characteristics.[87] There is a hydrophilic "head" and a hydrophobic "tail" to surfactants.[88,89] This lowers the interfacial tension and makes it easier for molecules with different polarities to interact by placing themselves at the interface. The surface tension or the interfacial tension relationship between two distinct phases affects many surfactant properties.[87] These characteristics are crucial for system characteristics, including the presence and persistence of emulsions or foams, which are discussed in ref. 90. With the help of structures called micelles, which are created by surfactants, hydrophilic and hydrophobic molecules can mix together in a single phase. These characteristics play a key role in the characteristics of systems, such as the presence and persistence of emulsions or foams, for which surfactants enable the mixing of hydrophilic and hydrophobic molecules by creating micelle-like

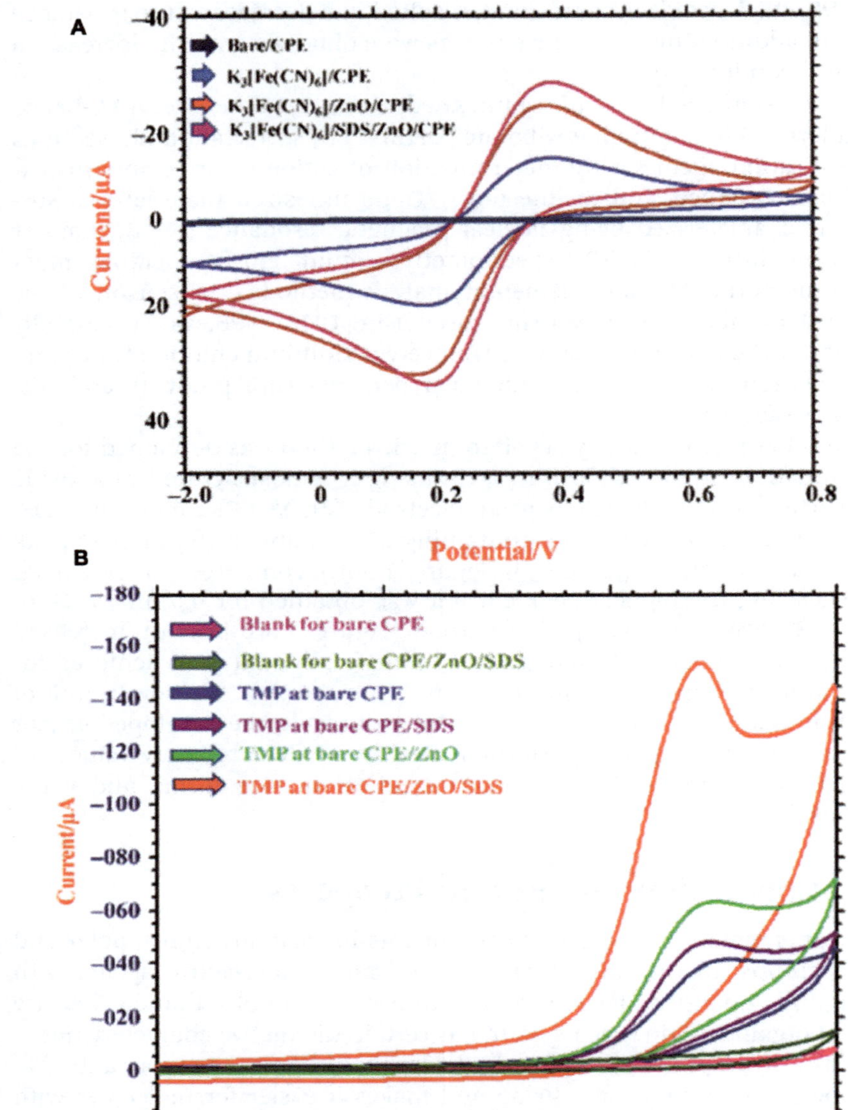

Figure 1.5 (A) CV measurements of 1.0 mM $K_3(Fe(CN)_6)$ at bare CPE, ZnO/CPE, and ZnO/SDS/CPE, (B) CV responses of 0.5 mM TMP in pH 3.0 (0.2 M PBS) at CPE, SDS/CPE, ZnO/CPE, and ZnO/SDS/CPE. Reproduced from ref. 26, https://doi.org/10.3390/bioengineering9100521, under the terms of CC BY 4.0 license, https://creativecommons.org/licenses/by/4.0/.

structures that enable both types of molecule to combine in a single phase. Codissolution, another name for this compatibility of molecules without a natural affinity, is useful for a variety of applications.[87] Their applicability is mostly determined by specific characteristics of the hydrophilic head.[88,89] Detergents, personal care products, lubricants, pharmaceuticals, bioremediation, and agriculture are just a few industries that commonly use surfactants. The most common classes of surfactants are anionic and non-ionic.[88,89,91,92]

The majority of modern surfactant manufacture is environmentally harmful and reliant on fossil fuels. Products may have troubling safety and health problems.[91] The sustainable manufacture of bio-based surfactants has important advantages in terms of biodegradability and biocompatibility, even though petrochemical synthesis is now more economically viable.[88,89] More importantly, consumer demands and political incentives may soon challenge the conventional wisdom that producing surfactants from renewable raw materials is just a means of enhancing agro-industrial side-streams and that the high consumption of surfactants proves their sustainable manufacture.[93] There is always a need to find surfactants that are less detrimental to the environment, and advancements in the creation of "green" surfactants are promising in terms of their contribution to "green analytical chemistry".[27]

The two main uses of surfactants in electrochemical analysis that are discussed in this chapter in electrolyte solutions and modified electrodes. Surfactants provide several advantages in electrochemical analysis such as improving the sensitivity of the sensor, increasing the analytical signal, and decreasing the limit of detection because of the numerous charged groups and micelles they contain. Surfactants also change the morphological properties of the nanomaterial, such as size and form, and are utilized in the production of nanomaterials.[27,63]

Electrochemical procedures are widely utilized because they are inexpensive, quick and simple to prepare, and produce outstanding analytical results. It is feasible to significantly increase sensitivity and selectivity by altering the redox potential, charge transfer coefficients, and electron transfer rate by adding surfactants to electrolytes or modifying the electrode surface. Various surfactant-based working electrode modifiers, such as composite materials, polymers, and nanoparticles, have also been discussed along with their applications in sensors. The sensitivity frequently increases as the length of the hydrophobic component of the surfactant increases unless the longer electron transfer distance to the target decreases the electron transfer rate. Superior electroanalytical properties, such as a high surface-to-volume ratio, powerful adsorption, high electrical conductivity, chemical stability, a wide potential range, a low background current, and acceptable biocompatibility, have been demonstrated for surfactant-modified electrodes.[63]

Electrochemical sensors based on surfactants can assess real samples in a variety of fields, including healthcare, food, agriculture, and industry. Due to their distinctive qualities, such as lower CMC, higher effectiveness in reducing the surface tension of water, stronger solubility, and greater capacity for surface wetting, surfactants facilitate electron transfer and exhibit various advantages over conventional surfactants. Consequently, they open up new possibilities for creating customized electrodes.[63]

Surfactants show the diversity of their effects in electrochemistry, catalysis of organic reactions, dissolution of organic compounds, control of electrochemical reactions, changing or increasing reaction rates, and controlling reaction pathways. Their use in electroplating techniques, corrosion prevention, improvements in the performance of batteries and fuel cells, and new or improved electroanalytical techniques show the diversity and importance of surfactants in research studies in these directions. It allows the visualization of possible structures of surfactants depending on experimental conditions.[17] We have discussed surfactants utilized as surface active agents, their characteristics, and their function in electrochemistry in this chapter.

Acknowledgements

The Council of Higher Education (YOK) was greatly appreciated for providing scholarships under the special 100/2000 scholarship program to Didem Nur Unal.

References

1. S. De, S. Malik, A. Ghosh, R. Saha and B. Saha, *RSC Adv.*, 2015, **5**, 65757–65767.
2. G. G. Ying, *Environ. Int.*, 2006, **32**, 417–431.
3. S. Kumar Shah, G. Chakraborty, A. Bhattarai and R. De, *J. Mol. Liq.*, 2022, **368**, 120678.
4. K. Alexander, S. Sheshrao Gajghate, A. Shankar Katarkar, A. Majumder and S. Bhaumik, *Mater. Today: Proc.*, 2020, **44**, 1136–1143.
5. D. Sivaganesh, S. Saravanakumar, V. Sivakumar, R. Rajajeyaganthan, M. Arunpandian, J. Nandha Gopal and T. K. Thirumalaisamy, *Mater. Charact.*, 2020, **159**, 110035.
6. M. Elmowafy, K. Shalaby, M. H. Elkomy, O. Awad Alsaidan, H. A. M. Gomaa, M. A. Abdelgawad, D. Massoud, A. Salama and K. M. El-Say, *Int. J. Pharm.*, 2022, **629**, 122375.
7. Z. Liu, A. B. Lansley, T. N. Duong, J. D. Smart and A. S. Pannala, *Biomolecules*, 2022, **12**, 1739.
8. M. Kumar, D. Elahi, A. Bhardwaj, S. Sharma, K. Khushi, E. Singh, N. Singh and A. Srivastava, *J. Mol. Liq.*, 2022, **364**, 119971.
9. S. M. Shaban, J. Kang and D. H. Kim, *Compos. Commun.*, 2020, **22**, 100537.
10. D. Mijaljica, F. Spada and I. P. Harrison, *Molecules*, 2022, **27**, 2010.
11. Y. Xie and X. He, *Colloids Surf., A*, 2023, **660**, 130850.

12. A. Elaraby, A. Elgendy, M. Abd-El-Raouf, M. A. Migahed, A. S. El-Tabei, A. M. Abdullah, N. H. Al-Qahtani, S. M. Alharbi, S. M. Shaban, D. Kim and N. M. El Basiony, *Colloids Surf., A*, 2023, **659**, 130687.
13. C. Verma, C. M. Hussain, M. A. Quraishi and A. Alfantazi, *Adv. Colloid Interface Sci.*, 2023, **311**, 102822.
14. A. A. Salman, *Carbohydr. Res.*, 2022, **520**, 108634.
15. P. Liu, J. Wu, X. Pei, Z. Cui, J. Jiang, B. Song and B. P. Binks, *Green Chem.*, 2022, **24**, 7612–7621.
16. O. T. Isaac, H. Pu, B. A. Oni and F. A. Samson, *Energy Rep.*, 2022, **8**, 2806–2830.
17. R. Vittal, H. Gomathi and K.-J. Kim, *Adv. Colloid Interface Sci.*, 2006, **119**, 55–68.
18. N. S. Prinith and J. G. Manjunatha, *Mater. Sci. Energy Technol.*, 2019, **2**, 408–416.
19. J. I. Gowda, S. T. Nandibewoor and A. T. Buddanavar, *Anal. Bioanal. Electrochem.*, 2016, **8**, 329–344.
20. S. A. Hoshyar, H. A. H. Barzani, Y. Yardım and Z. Şentürk, *Colloids Surf., A*, 2021, **610**, 125916.
21. K. Fic, G. Lota and E. Frackowiak, *Electrochim. Acta*, 2010, **55**, 7484–7488.
22. S. Afroukhteh, C. Dehghanian and M. Emamy, *Prog. Nat. Sci.: Mater. Int.*, 2012, **22**, 480–487.
23. M. A. Deyab, S. S. A. El-Rehim and S. T. Keera, *Colloids Surf., A*, 2009, **348**, 170–176.
24. A. S. Fouda, M. A. A. El-Ghaffar, M. H. Sherif, A. T. El-Habab and A. El-Hossiany, *Prot. Met. Phys. Chem. Surf.*, 2020, **56**, 189–201.
25. S. M. Tawfik and N. A. Negm, *J. Mol. Liq.*, 2016, **215**, 185–196.
26. V. B. Patil, D. Ilager, S. M. Tuwar, K. Mondal and N. P. Shetti, *Bioengineering*, 2022, **9**(10), 521.
27. D. N. Unal, S. Yıldırım, S. Kurbanoglu and B. Uslu, *TrAC, Trends Anal. Chem.*, 2021, **144**, 116418.
28. J. Kozak, K. Tyszczuk-Rotko, K. Sztanke and M. Sztanke, *Int. J. Mol. Sci.*, 2023, **24**(1), 564.
29. C. E. Sener, B. Dogan Topal and S. A. Ozkan, *Anal. Bioanal. Chem.*, 2020, **412**, 8073–8081.
30. O. Selcuk, C. Erkmen, B. Bozal-Palabiyik and B. Uslu, *Electroanalysis*, 2021, **33**, 1290–1298.
31. W. N. Maclay, *J. Colloid Sci.*, 1956, **11**, 272–285.
32. M. Moradi and Y. Yamini, *J. Sep. Sci.*, 2012, **35**, 2319–2340.
33. H. Cortés, H. Hernández-Parra, S. A. Bernal-Chávez, M. L. Del Prado-Audelo, I. H. Caballero-Florán, F. V Borbolla-Jiménez, M. González-Torres, J. J. Magaña and G. Leyva-Gómez, *Materials*, 2021, **14**(12), 3197.
34. E. Fuguet, C. Ràfols, M. Rosés and E. Bosch, *Anal. Chim. Acta*, 2005, **548**, 95–100.
35. K. Nesměrák and I. Němcová, *Anal. Lett.*, 2006, **39**, 1023–1040.
36. M. M. Mabrouk, N. A. Hamed and F. R. Mansour, *Appl. Spectrosc. Rev.*, 2021, 1–29.

37. D. Otzen, *Biochim. Biophys. Acta, Proteins Proteomics*, 2011, **1814**, 562–591.
38. G. Kume, M. Gallotti and G. Nunes, *J. Surfactants Deterg.*, 2008, **11**, 1–11.
39. F. Liu, W. Yuan, T. Li and Z. Zhang, *Mater. Res. Innovations*, 2015, **19**, S2-70–S2-75.
40. P. A. Pushpanjali, J. G. Manjunatha and M. T. Shreenivas, *ChemistrySelect*, 2019, **4**, 13427–13433.
41. G. Tigari, J. G. Manjunatha, C. Raril and N. Hareesha, *ChemistrySelect*, 2019, **4**, 2168–2173.
42. C. Xu, G. Xu, Y. Zhang, S. Fang, P. Nie, L. Wu and X. Zhang, *ACS Energy Lett.*, 2017, **2**, 2659–2666.
43. X. Xing, R. Liu, S. Liu, S. Xiao, Y. Xu, C. Wang and D. Wu, *Electrochim. Acta*, 2016, **194**, 310–316.
44. C. Zhou and Y. Wang, *Curr. Opin. Colloid Interface Sci.*, 2020, **45**, 28–43.
45. S. P. Moulik, M. E. Haque, P. K. Jana and A. R. Das, *J. Phys. Chem.*, 1996, **100**, 701–708.
46. S. Sarkar, A. Maitra, S. Banerjee, V. S. Thoi and J. M. Dawlaty, *J. Phys. Chem. B*, 2020, **124**, 1311–1321.
47. S. Banerjee, Z.-Q. Zhang, A. S. Hall and V. S. Thoi, *ACS Catal.*, 2020, **10**, 9907–9914.
48. S. Devaraj and N. Munichandraiah, *J. Electrochem. Soc.*, 2007, **154**, A901.
49. M. Ghaemi, E. Ghafouri and J. Neshati, *J. Power Sources*, 2006, **157**, 550–562.
50. M. Kawakami, K. Tanaka, N. Uriuda and S. Gondo, *Bioelectrochemistry*, 2000, **52**, 51–56.
51. R. Aslam, M. Mobin, J. Aslam, A. Aslam, S. Zehra and S. Masroor, *Adv. Colloid Interface Sci.*, 2021, **295**, 102481.
52. N. Parris, C. Pierce and W. M. Linfield, *J. Am. Oil Chem. Soc.*, 1977, **54**, 294–296.
53. R. Sarkar, A. Pal, A. Rakshit and B. Saha, *J. Surfactants Deterg.*, 2021, **24**, 709–730.
54. A. P. Gerola, P. F. A. Costa, F. Nome and F. Quina, *Curr. Opin. Colloid Interface Sci.*, 2017, **32**, 48–56.
55. C. Puri, L. Pucciarini, M. Tiecco, V. Brighenti, C. Volpi, M. Gargaro, R. Germani, F. Pellati, R. Sardella and C. Clementi, *Antioxidants*, 2020, **9**(11), 1055.
56. A. S. Burlachenko, O. V. Salishcheva, L. S. Dyshlyuk and A. Y. Prosekov, *Appl. Sci.*, 2021, **11**, 8939.
57. T. M. McCoy, J. B. Marlow, A. J. Armstrong, A. J. Clulow, C. J. Garvey, M. Manohar, T. A. Darwish, B. J. Boyd, A. F. Routh and R. F. Tabor, *Biomacromolecules*, 2020, **21**, 4569–4576.
58. Y. Zhu, X. Li, D. Zhang and L. Gao, *J. Power Sources*, 2021, **515**, 230646.
59. R. Fuchs-Godec, *Electrochim. Acta*, 2009, **54**, 2171–2179.
60. J. H. Lee, J. C. Shin, J. Kim, J.-W. Ho, W. J. Cho, M. J. Park, G.-R. Yi, M. Lee and P. J. Yoo, *J. Power Sources*, 2023, **557**, 232565.

61. N. O. Decarli, E. Zapp, B. S. de Souza, E. R. Santana, J. P. Winiarski and I. C. Vieira, *Biochem. Eng. J.*, 2022, **186**, 108565.
62. H. Li, Z. Deng, Q. Tian, L. Lun, P. Zhao, X. Yang, J. Shen, B. Jiang, Y. Zhou and T. Zhou, *Int. J. Environ. Anal. Chem.*, 2020, 1–14.
63. M. R. Housaindokht, F. Janati-Fard and N. Ashraf, *J. Surfactants Deterg.*, 2021, **24**, 873–895.
64. G. Soberón-Chávez and R. M. Maier, in *Biosurfactants*, ed. G. Soberón-Chávez, Springer Berlin Heidelberg, Berlin, Heidelberg, 2011, pp. 1–11.
65. A. R. Markande, D. Patel and S. Varjani, *Bioresour. Technol.*, 2021, **330**, 124963.
66. G. K. Ziyatdinova, E. R. Ziganshina and H. C. Budnikov, *J. Anal. Chem.*, 2012, **67**, 869–879.
67. L. Shulman, L. Pei, M. F. Bahnasy and C. A. Lucy, *Analyst*, 2017, **142**, 2145–2151.
68. G. Płaza and V. Achal, *Int. J. Mol. Sci.*, 2020, **21**, 2152.
69. S. S. Siwal, Q. Zhang, A. K. Saini, V. K. Gupta, D. Roberts, V. Saini, F. Coulon, B. Pareek and V. K. Thakur, *J. Environ. Chem. Eng.*, 2021, **9**, 105982.
70. H. Yan, M. Sone, A. Mizushima, T. Nagai, K. Abe, S. Ichihara and S. Miyata, *Surf. Coat. Technol.*, 2004, **187**, 86–92.
71. J. Lv, X. Zhao, X. Jie, J. Li, X. Wei, B. Chen, G. Hong, W. Wu and L. Wang, *ChemElectroChem*, 2019, **6**, 3254–3263.
72. Y.-J. Shih and Z.-L. Wu, *Appl. Catal., B*, 2021, **285**, 119784.
73. G. S. Frankel, in *Active Protective Coatings New-Generation Coatings for Metals*, ed. A. E. Hughes, J. M. C. Mol, M. L. Zheludkevich and R. G. Buchheit, Springer Netherlands, Dordrecht, 2016, pp. 17–32.
74. M. A. Bedair, M. M. B. El-Sabbah, A. S. Fouda and H. M. Elaryian, *Corros. Sci.*, 2017, **128**, 54–72.
75. S. M. Shaban, A. A. Abd-Elaal and S. M. Tawfik, *J. Mol. Liq.*, 2016, **216**, 392–400.
76. A. Abd-ElHamid, W. El-dougdoug, S. M. Syam, I. Aiad, S. M. Shaban and D.-H. Kim, *J. Mol. Liq.*, 2023, **369**, 120890.
77. R. Jalab, M. A. Saad, M. H. Sliem, A. M. Abdullah and I. A. Hussein, *Molecules*, 2022, **27**, 6414.
78. M. S. Whittingham, R. F. Savinell and T. Zawodzinski, *Chem. Rev.*, 2004, **104**, 4243–4244.
79. S. Yang, J. Yang, Y. Chung and Y. Kwon, *Int. J. Hydrogen Energy*, 2017, **42**, 17211–17220.
80. F. Meng, C. Sun, J. Shi, H. Zhang, B. Xu and Y. Ding, *Int. J. Hydrogen Energy*, 2019, **44**, 1122–1129.
81. B. Begum, S. Bilal, A. U. Shah and P. Röse, *Polymers*, 2021, **13**, 2883.
82. Y. Xue and Y. Wang, *Green Chem.*, 2020, **22**, 6288–6309.
83. J. Shu, F. Lin, M. Chen, B. Li, L. Wei, J. Wang, Z. Luo and R. Wang, *Hydrometallurgy*, 2020, **193**, 105311.
84. Y. Tian, J. Shu, M. Chen, J. Wang, Y. Wang, Z. Luo, R. Wang, F. Yang, F. Xiu and Z. Sun, *J. Cleaner Prod.*, 2019, **236**, 117708.

85. A. Nag, N. Afsarimanesh, S. Nuthalapati and M. E. Altinsoy, *Materials*, 2022, **15**, 4504.
86. N. Sakač, D. Marković, B. Šarkanj, D. Madunić-Čačić, K. Hajdek, B. Smoljan and M. Jozanović, *Molecules*, 2021, **26**, 3627.
87. S. Ebnesajjad, *Surface Treatment of Materials for Adhesive Bonding*, Elsevier, 2014.
88. L. Wang and Y. Queneau, *Encycl. Sustainability Sci. Technol.*, 2018, p. 1.
89. L. A. Sarubbo, M. da G. C. Silva, I. J. B. Durval, K. G. O. Bezerra, B. G. Ribeiro, I. A. Silva, M. S. Twigg and I. M. Banat, *Biochem. Eng. J.*, 2022, **181**, 108377.
90. M. Poša, A. Pilipović, L. Torović and J. C. Hogervorst, *J. Mol. Liq.*, 2019, **273**, 134–146.
91. P. G. Jessop, F. Ahmadpour, M. A. Buczynski, T. J. Burns, N. B. Green Ii, R. Korwin, D. Long, S. K. Massad, J. B. Manley and N. Omidbakhsh, *Green Chem.*, 2015, **17**, 2664–2678.
92. A. Bhadani, A. Kafle, T. Ogura, M. Akamatsu, K. Sakai, H. Sakai and M. Abe, *Curr. Opin. Colloid Interface Sci.*, 2020, **45**, 124–135.
93. J. W. Agger and B. Zeuner, *Curr. Opin. Biotechnol.*, 2022, **78**, 102842.

CHAPTER 2

New Approach to the Preparation of Cationic, Anionic, Neutral, and Gemini Sensors

MOHAN KUMAR,*[a] H. NAGARAJAIAH,[b] VINUTH MIRLE,[c] R. S. VEERAPUR,[d] PRASAD NAGARAJ BAPAT,[e] PRAMOD GOPAL PAI[e] AND J. G. MANJUNATHA[f]

[a] Department of Chemistry, PES Institute of Technology and Management, Shivamogga, 577 204, Karnataka, India; [b] Department of Chemistry, School of Applied Science, REVA University, Yelahanka, 560064, Bangalore, Karnataka, India; [c] Department of Chemistry, The National Institute of Engineering (NIE), North Campus, Koorgalli-570018, Mysore, Karnataka, India; [d] Department of Metallurgy & Materials Engineering, Malawi Institute of Technology – Malawi University of Science and Technology, P.O. Box 5196, Limbe, Malawi; [e] Department of Physics, PES Institute of Technology and Management, Shivamogga, 577 204, Karnataka, India; [f] Department of Chemistry, FMKMC College, Constituent College of Mangalore University, Madikeri, Karnataka, India
*Email: mohankumar.mogi@gmail.com

2.1 Introduction

Surfactants are surface-active agents, which are amphiphilic molecules containing hydrophobic and hydrophilic moieties.[1] They consist of a head and a tail; surfactants enclosing a polar head can be ionic or non-ionic, and these can

Detection Science Series No. 23
Surfactant-based Sensors in Chemical and Biochemical Detection
Edited by J. G. Manjunatha

Published by the Royal Society of Chemistry, www.rsc.org

also be divided into anionic (examples: sodium dodecylbenzenesulfonate and sodium dodecylsulfate (SDS)), cationic (examples: cetyltrimethylammonium bromide (CTAB), *N*-dodecylpyridinium bromide, and cetylpyridinium bromide (CPB)), and zwitterionic or amphoteric (example: phosphatidylcholine) surfactants.[2] Generally, non-ionic surfactants (examples: Tween 20 and Triton X-100) contain a polar head. Negatively and positively charged surfaces are made by anionic and cationic surfactants, respectively. However, amphoteric/zwitterionic surfactants produce negative or positive charges depending on the solution's pH value.[3] Moreover, gemini surfactants come under novel surfactants and comprise more than one hydrophobic tail group and hydrophilic head group connected by a spacer near or at the head group.[4–6] Figure 2.1 represents anionic, cationic, non-ionic, zwitterionic, and gemini surfactants.

A sensor is a device that detects and transforms a variable value into a recognized electronic signal. The more significant norms for estimating the performance of a sensor are its stability, selectivity and sensitivity.[7] Normally, methods such as fluorescence spectrometry, spectrophotometry, colorimetry, UV-visible spectroscopy, and electrochemical techniques are used in sensor design.[8–13] However, these methods, excluding electrochemical techniques may need a lengthy sample preparation and make analytes of lower concentration difficult to determine because of insufficient sensitivity and matrix interference. Besides, there is a requirement for professional training. To overcome these limitations, attention has been given to the capability to reach more selectivity and sensitivity by electrochemical techniques. Nowadays, electrochemical sensors are used as a vital kind of chemical sensor that has various advantages, including fast response, low detection limit, absolute discrimination capability, high accuracy, and appropriate field-based size.[14,15] Electrochemical sensors are generally operated using a three-electrode system

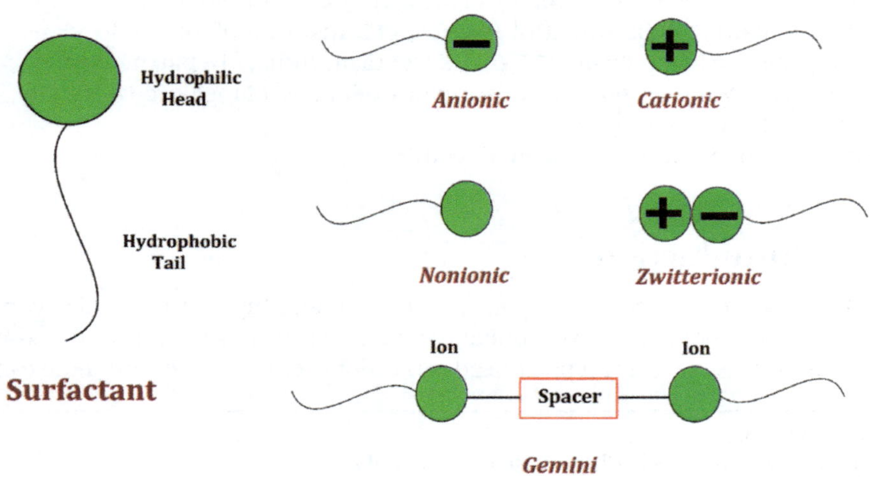

Figure 2.1 Schematic representation of anionic, cationic, non-ionic, zwitterionic, and gemini surfactants.

(counter, reference, and working electrodes; usually, working electrodes can be modified by different materials or by different methods). These sensors have obtained great attention in practical as well as fundamental research applications, due to the various above-mentioned benefits. In electrochemical techniques, voltammetry is a significant method on which various electrochemical sensors depend. Here, the analyte can be simply determined from the voltammetric peak potential. Moreover, modified voltammetric electrodes have been broadly utilized as sensors to confirm high operational and analytical characteristics for analyte determination.[16,17]

Presently, much attention has been paid to utilizing surfactants for the fabrication and modification of electrodes. Several results have been obtained with adsorbed surfactants on the electrode surface, such as higher electron transfer rate, changes in the redox potential and charge transfer coefficients, improving their potential to amplify signals.[18,19] Moreover, surfactants enclose a broad range of properties that make them a feasible preference for checking selective features of the response of the sensor towards the analytes. This consequence is due to the capability of surfactants to alter the surface of electrodes in terms of lipophilic, hydrophilic and electrostatic features.[20,21]

2.2 New and Sustainable Approach to the Preparation of Electrochemical Sensors

The research field has been focused on developing a different variety of electrode modifications. Chemical modification creates highly active redox sites that transport electrons between the electrode substrate and an analyte solution and significantly reduce the activation of overpotential. Therefore, chemically modified electrodes (CMEs) offer significantly improved electrocatalytic activity towards the determination of various biomolecules, environmental hazards, *etc.* because of certain unique characteristics of the materials used for modification.[22] Electrodes fabricated by chemical modification are less prone to oxide formation and surface fouling characteristics compared to unmodified electrodes. Chemical modifiers include, but are not limited to, conducting polymers (CPs), nanomaterials, inorganic complexes, ionic liquids (ILs), composite materials, and surfactants.[23–28] As far as trends in merging various active materials modification are concerned, they are a combination of relatively familiar materials such as metals or their oxides, nanoparticles, carbon, IL, CPs, *etc.*[22]

Recently, many researchers have utilized electrochemical sensors by modifying electrodes with surfactants for the determination of biomolecules, environmentally hazardous molecules and organic species.[29–32] Shankar *et al.* prepared a carbon paste electrode (CPE) by immobilizing the CTAB surfactant on the electrode surface, which was utilized for the detection of dopamine (DA), ascorbic acid (AA), and uric acid (UA), and the study of some electrochemical parameters.[33] Hosseinzadeh *et al.* examined the effect of a cationic surfactant (CTAB) on the determination of DA and AA by modifying a CPE with tin hexacyanoferrate, and concluded that the CTAB functioned as a strong electrocatalyst towards AA oxidation, and also avoided the overlap of DA and AA anodic peaks. Generally, an immobilization method is used to modify the electrode surface using surfactants. In one research work,[34] ZnO was prepared by a solution-free mechanochemical method and the CPE was modified using prepared ZnO. The as-prepared electrode was further modified using an immobilizing surfactant (Tx-100/CTAB) on the electrode surface (Figure 2.2(a)). The surfactant-based ZnO-modified CPE was employed as an electrochemical biosensor for the determination of the biomolecule dopamine (Figure 2.2(b)). As shown in the figure, the DA undergoes electrochemical oxidation on the surface of the surfactant-based electrode.[35]

Vernekar *et al.* have developed a sensor using glucose-modified CPE for the highly sensitive detection of a non-steroidal anti-inflammatory drug (mefenamic acid) in the presence of CTAB.[36] Kumar *et al.* prepared the modified CPE using a Co_3O_4/CuO composite nanopowder and immobilized the SDS for the sensitive and simultaneous determination of AA, DA, and UA.[37] El-Said *et al.* used gemini surfactant for the electrochemical determination of As(III) at the indium tin oxide electrode. Moreover, they compared the electrochemical response of the gemini surfactant-modified electrode with a further change to the experimental setup in the presence of CTAB and other surfactants.[38] Tigari *et al.* fabricated a carbon nanotube paste electrode modified with sodium lauryl sulfate (SLS) for the electrochemical detection of riboflavin and studied the reproducibility, repeatability, stability, and sensitivity of the modified carbon nanotube paste electrode.[39] Alanine and tyrosine are biomolecules used for the functionalization of graphene oxide (GO) nanoflakes, which were utilized for the modification of CPE, and these electrodes were further modified by immobilizing sodium alpha olefin sulfonate (SAOS) surfactant. Their performance for the simultaneous detection of DA and AA was examined; alanine-functionalized GO nanoflake-modified SAOS CPE showed enhanced performance compared to other modified electrodes in their studies.[40] Zahid *et al.* fabricated an ultrasensitive electrochemical sensor using a urea

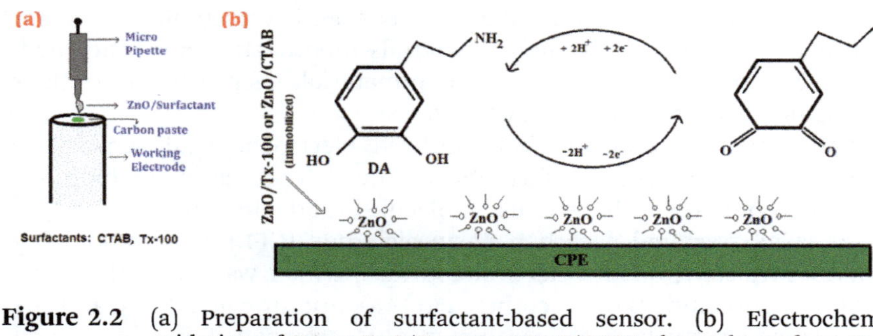

Figure 2.2 (a) Preparation of surfactant-based sensor. (b) Electrochemical oxidation of DA at ZnO/Tx-100 or ZnO/CTAB electrode surface. Reproduced from ref. 35 with permission from The Royal Society of Chemistry.

functionality-based surfactant, 1-(2,4-dinitrophenyl)-dodecanoylthiourea immobilized on a glassy carbon electrode for the successful detection of mercury.[41] Most of the literature has concluded that electrochemical sensors have more advantages compared to various other types of sensors because of their low detection limits, high accuracy, fast response, absolute discrimination capability, low cost, and field-based size.[42–45]

2.3 Types of Surfactants

Surfactants are recognized by having two types of molecular structures, *i.e.* hydrophilic and hydrophobic (lipophilic). The hydrophilic portion is a polar group containing functional groups such as alcohol, acid, ester, amine, amide, thiol, ether, sulfate, phosphate, and sulfonate having dynamic affinity to dissolve in polar media like water. The hydrophobic group is the non-polar part of surfactants which includes halogen, aromatic hydrocarbon, paraffin, and cycloparaffin with a strong affinity for non-polar media like oils and fats. Hence, the surfactants can interact with polar and non-polar media simultaneously. Knowing how to classify surfactants will help us to understand how to use them in analytical approaches. According to their hydrophilic head and the source from which they are made, surfactants can be categorized.[46]

2.3.1 Classification Based on Hydrophilic Head Parts

2.3.1.1 Anionic Surfactants

Anionic surfactants are those in which the hydrophilic heads are negatively charged. Because of the negatively charged groups, anionic surfactants can be complexed with the oppositely charged ions in water/solution. Anionic groups are directly connected to the hydrophobic part through ether, ester, amide or amidine linkages.[47] Carboxylic acid- and sulfonic acid-based anionic surfactants are the most widely utilized (Figure 2.3). Among them are fatty acids (found in soaps), alkylbenzenesulfonates (found in detergents), lauryl sulfate (found in foaming agents), di-alkyl sulfosuccinate (found in wetting agents), and lignosulfonates (found in dispersants). Fats from plants and animals that exist naturally are the source of aliphatic carboxylic acid-based surfactants. Typically, they are found in nature as triesters of carboxylic acids and glycerol. These fats can be hydrolyzed to soaps through the saponification process using potassium/sodium hydroxide. Carboxylic acids are obtained in the form of sodium or potassium salts, also referred as fatty acids, which makes them anionic surfactants. Some examples in this category include sodium stearate, a sodium salt of steric acid obtained by saponification of animal fat; this is the most common soap. It is also one of the main components of food additives and flavorings.[48] Other carboxylic acid-based surfactants include sodium lauroylsarcosinate,[49] sodium lauryl sarcosinate, glycolic acid ethylate-4-*tert*-butylphenyl ether and glycolic acid ethylate laurylphenyl ether.[50]

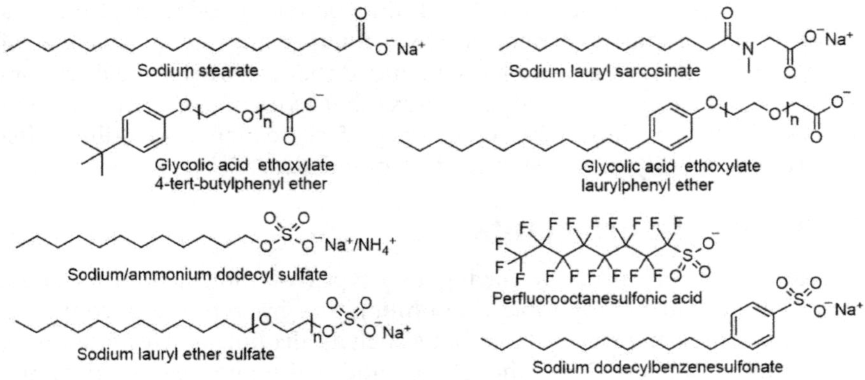

Figure 2.3 Anionic surfactants based on sulfonic and carboxylic acids.

Sulphonic acid-based surfactants are also an important class of anionic surfactants. Among many anionic surfactants, sodium/ammonium dodecyl sulfate is the most important one, and is used in the denaturation of proteins.[51] In addition to aliphatic sulfonic acids (perfluorooctane sulfonic acid and sodium lauryl ether sulfate) many aromatic sulfonic acid surfactants (commonly used is sodium dodecylbenzenesulfonate) are used as sanitizers, and added to fruit and vegetable wash water. They affect the performance of wash water by means of the removal of surface bacteria, reducing the transfer of planktonic bacteria and reducing the risk of cross-contamination.

2.3.1.2 Cationic Surfactants

Cationic surfactants with hydrophilic heads are positively charged. These surfactants' positive charge can be temporary or limited to specific pH ranges. Cationic surfactants rely mostly on nitrogen-containing functional groups such as amine, quaternary ammonium, imidazolium, and pyridinium. The amine surfactants are active only at low pH values, hence, these can potentially be used to make pH sensors. Cationic surfactants stick to most surfaces, creating self-assembled monolayers. Therefore, they are preferentially used in softeners and hair conditioners, or as anticorrosion agents and antistatic agents. Quaternary ammonium surfactants also have an antibacterial effect and can be used as preservatives. Cationic surfactants are also used in the heavy-duty cleaners used in carpet cleaners. They have been used in various fields such as corrosion inhibitors, emulsifiers, fabric softeners, particle dispersants, and rinse aids.[52–56]

Generally used quaternary ammonium compounds include aliphatic (cetylpyridinium chloride, cetrimonium bromide, dimethyldioctadecylammonium chloride, *etc.*), and aromatic (alkylbenzenes such as benzethonium chloride and pyridine derivatives) (Figure 2.4). The most popular cationic surfactant cetyltrimethylammonium bromide (CTAB) is used in electrochemical sensors for the detection of various organic molecules and biospecies.[57]

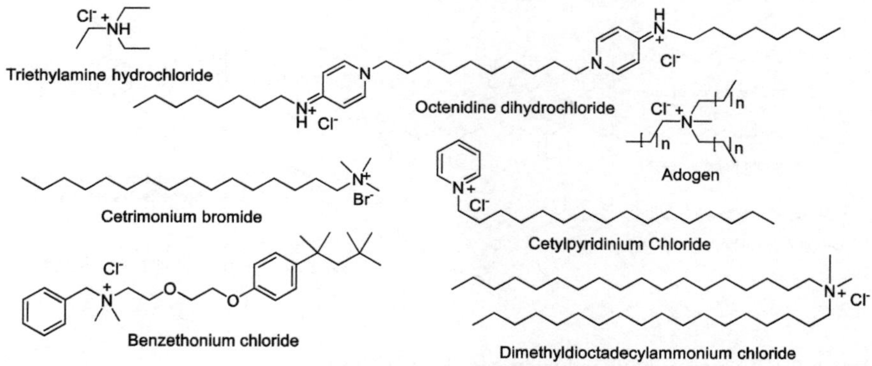

Figure 2.4 Cationic surfactants that are frequently utilized.

In general, cationic surfactants cost more than anionic ones, because a high-pressure hydrogenation reaction must take place during synthesis. Since there is no less expensive alternative, they are only used in two situations: as bactericides and as positively charged substances that can adsorb on negatively charged surfaces to produce an antistatic and hydrophobic effect, which is frequently of great commercial importance, such as in corrosion inhibition.[58]

2.3.1.3 Zwitterionic Surfactants

Surfactants that are zwitterionic or amphoteric display both cationic and anionic characteristics. Depending on the structure of amphoteric surfactants, the anionic components are typically carboxylic, sulfuric, or phosphoric acids, whereas the cationic components are typically amine or quaternary ammonium cations (or esters). Depending on the pH, they exhibit anionic or cationic behavior. In comparison, the anionic character predominates in the local surroundings, the cationic character predominates in acidic conditions.[59] Examples include artificial chemicals like betaines or organic compounds like saponins, sulfobetaines, amino acids, and phospholipids. Amphoteric surfactants are often highly expensive, which limits their use to very niche applications like cosmetics, where their excellent biological compatibility and low toxicity are of utmost significance.

Using quaternary ammonium with an attached hydroxy group will also provide zwitterionic surfactants. Although the hydroxy group is often just mildly acidic, if it is connected directly to a nitrogen atom, it becomes substantially more susceptible to deprotonation. Lauryl dimethylamine-*N*-oxide is a typical representative of this group of surfactants.

Phospholipids, which are made up of a quaternary ammonium head connected to two hydrophobic fatty acids by a phosphate group and glycerol, are a significant class of zwitterionic surfactants. These surfactants are among the most crucial in cell biology since they make up the membranes of cells.[60] While the anionic component is typically carboxylic or sulfuric, the

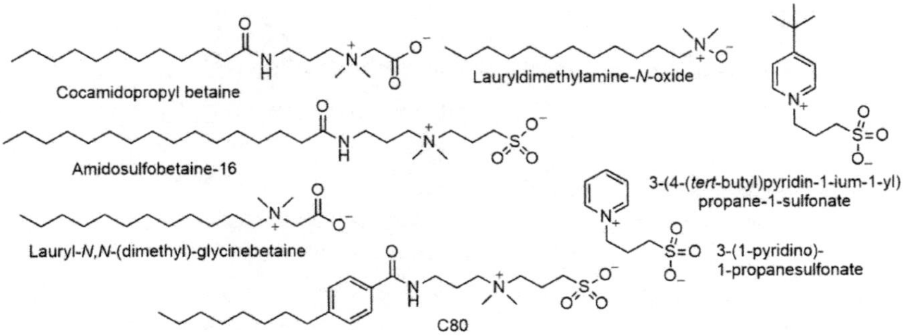

Figure 2.5 Commonly used zwitterionic surfactants.

cationic ingredient is always either an amine or a quaternary ammonium cation (Figure 2.5). Cocamidopropylbetaine, amidosulfobetaine-16, and C80 detergent are examples of a mixture of carboxylic acid with a quaternary ammonium ion. Lauryl-*N*,*N*-(dimethyl)-glycinebetaine is one of many different carboxylic acid/quaternary ammonium surfactants in use. It is common to utilize 3-(4-*tert*-butyl-1-pyridinio)-1-propanesulfonate and 3-(1-pyridinio)-1-propanesulfonate as aromatic quaternary ammonium surfactants (Figure 2.5).

2.3.1.4 Neutral Surfactants

Surfactants with non-ionic polar head groups are known as neutral surfactants. When in an aqueous solution, they don't ionize. These are non-dissociable hydrophilic groups, such as alcohol, phenol, ether, ester, or amide. Alcohols are generally strong candidates for possessing such head groups. Alcohols are moderately soluble in aqueous solutions and are weak acids.

Primary alcoholic groups bound to non-polar aliphatic tails, and cyclic, or aromatic compounds are typically well-suited for neutral surfactants. These neutral surfactants are made hydrophilic to a significant extent by the inclusion of a polyethylene glycol chain, which is produced by polycondensing ethylene oxide.

They often rely on a functional group with relatively little deprotonation ability. Non-ionic surfactants generally have lower solubility than ionic surfactants, but they don't affect the pH of the solution. The fact that their critical micelle concentrations (CMCs) are often higher than those of ionic surfactants may be advantageous because it enables the addition of additional surfactants to a solution without the danger of unintended micelle production. Neutral surfactants are crucial components of cell biology because they are gentle, do not disrupt protein–protein interactions, and have the ability to dissolve non-polar substances like bilayer membranes.[61] Octylglucoside is one of many non-ionic surfactants that are widely employed to solubilize integral membrane proteins for biochemical research.[62] In shampoos and creams, nonoxynol-9, which contains a hydrophobic *n*-nonane tail, has been utilized as a component. One alkylether of glycerol is glycerol monolaurate, which has a small amount of long-chain fatty acid esterification (lauric acid). As a surfactant, ethanolamine has a number of derivatives.

Typically, amides are produced by mixing ethanolamine and lauric acid to form a secondary amine surfactant, commonly called a cocamide surfactant. Lauramide monoethanolamine and lauramide diethanolamine are popular examples. Non-ionic surfactants that contain alkyl ethers of poly(ethylene glycol) are also often utilized. They are made up of an aliphatic hydrocarbon chains connected to poly(ethylene glycol) chains of various lengths. Penta/octaethylene glycol monododecyl ether is a good example of this surfactant family (Figure 2.6).

2.3.1.5 Gemini Surfactants

Two hydrophilic head groups and two hydrophobic tails are joined chemically by a spacer head group to form a gemini surfactant. The two polar head groups can be anionic, cationic, or non-ionic, while the terminal two hydrocarbon tails might be long or short. The spacer might be rigid or flexible, long or short.[63] When the hydrophobic and hydrophilic groups of a gemini surfactant are identical, a symmetrical structure is created (Figure 2.7). Gemini surfactants are superior in surface activity to conventional surfactants and can self-assemble at considerably lower concentrations. For adsorption and catalysis applications, analytical separations, gene transfection, improved oil recovery, biotechnology, nanoscale technology, paint additives, and solubilization procedures, these surfactants are particularly appealing. The demand for personal care items, household cleaners, laundry detergents, and industrial anticorrosion chemicals has led to an increase in the use of surfactants. In the near future, it is anticipated that this will result in the release of more novel and effective surfactant-based products. Gemini surfactants typically have CMCs that are hundreds of times lower than similar monomeric surfactants. Dimeric surfactants outperform their monomeric counterparts in terms of their ability to reduce

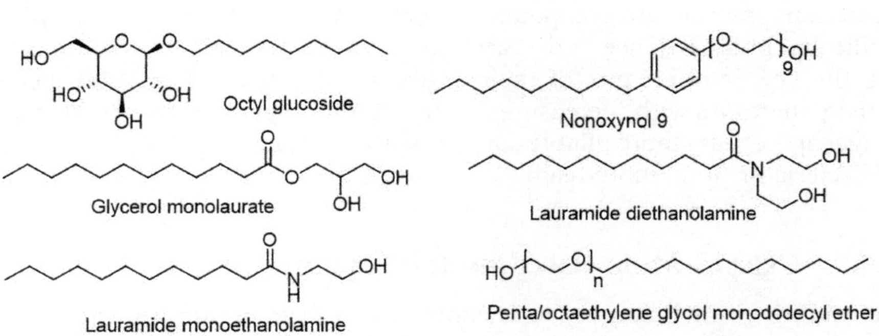

Figure 2.6 Commonly used non-ionic surfactants.

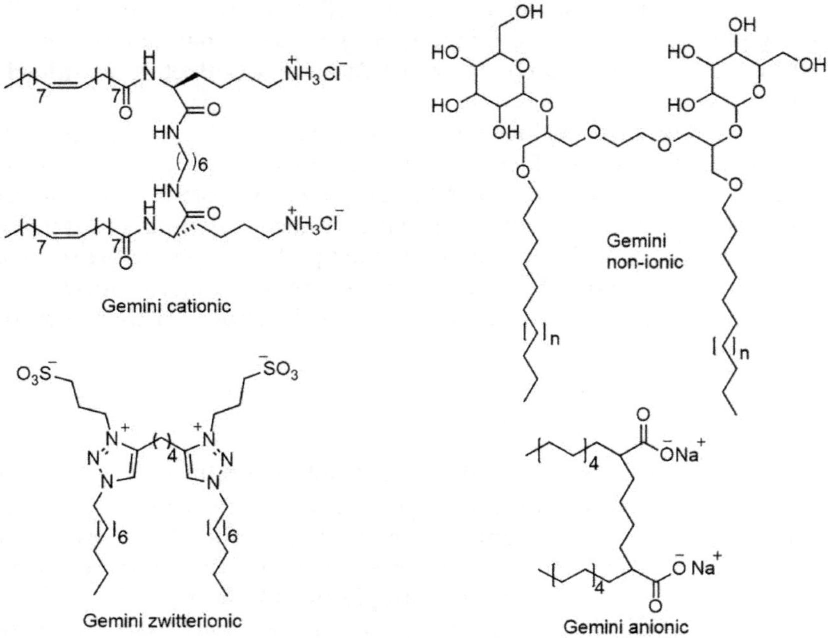

Figure 2.7 Representative examples of gemini (cationic, non-ionic, zwitterionic and anionic) surfactants.

surface tension. CMC decreases surface tension, and effective emulgation and solubilization effects, which all contribute to the surfactants' increased efficacy.

In an aqueous solution, the gemini alkylammonium salts exhibit remarkable surface and interfacial characteristics. Moreover, these surfactants are capable of forming a variety of morphological shapes in solution, including spherical, rod-shaped, elliptical, and worm-like micelles. Gemini surfactants' special qualities are used in dispersants, emulsifiers, coating agents, and corrosion inhibitors. It has been discovered that dimeric quaternary ammonium compounds are effective microbiocides. The hydrophilic–lipophilic balance and spacer length of quaternary ammonium salts are the key determinants of their antibacterial activity. The antibacterial activity increases with increasing distance because gemini surfactants with longer spacers are more pliable and link with the negatively charged surfaces of bacteria or fungi more readily.

2.3.2 Classification Based on the Source

Depending on their source, surfactants can be divided into two categories: synthetic and biosurfactants. Oleo-chemicals are made from vegetable oils and fats that have undergone a chemical transformation.[64]

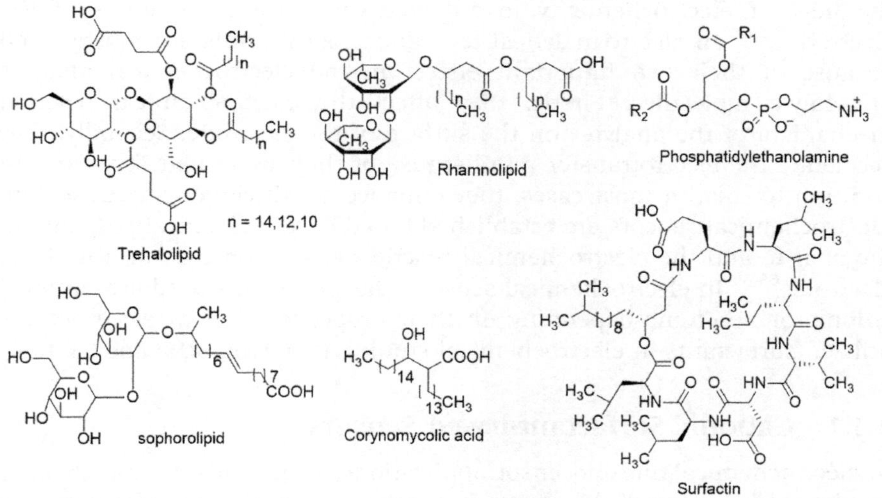

Figure 2.8 Different types of microbial surfactants.

2.3.2.1 Synthetic Surfactants

Chemical reactions are used to create synthetic surfactants, which mostly consist of ethylene and propylene oxide copolymers, alkoxylated alkylphenol derivatives, alcohols, sorbitan esters, and amines.[65] Polymeric surfactants, a new class of surface-active material recently discovered, are produced by joining one or more macromolecules with hydrophilic and lipophilic properties, either as distinct blocks or as grafts. They are currently widely employed in the creation of a variety of items, including cosmetics, paints, meals, and additives for petroleum production.

2.3.2.2 Biosurfactants

These are produced from biological sources. Depending on where they came from, they are separated into first-generation and second-generation. First-generation biosurfactants are derived from basic ingredients derived from plants and animals that are created by chemical synthesis. Second-generation biosurfactants are produced from biological processes such as the use of biocatalysts or by fermentation processes.[66] A few microbial surfactants are shown in Figure 2.8.

2.4 Preparation of Surfactant-based Sensors

Presently, the usage of surfactants is very necessary for various industries, particularly in analytical chemistry. Because of their unique structures, surfactants can be utilized appropriately in various fields such as the synthesis of nanomaterials to control their size and stability, in the preparation of samples using less organic solvent in chromatographic methods, and in the field of electrochemistry to enhance the surface properties of the electrodes.[1,67] In electroanalytical techniques, surfactants are mostly used because of their two important effects in the electrochemical analysis of analyte molecules. Firstly, they affect the electrochemical reaction mechanism of the analytes on the surface of the electrode. Secondly, they accelerate the electrotransfer rate because of their molecular structure. In addition to this, in most cases, they enhance sensitivity during detection. Electrochemical sensors are established based on the chemical structure of the analyte and the electrochemical reaction arising on the surface of the electrode.[68,69] In electrochemical sensors, the surfactants used are cationic, anionic or non-ionic depending on their properties. The most commonly utilized surfactants in electrochemical studies are CTAB, SDS, and Tx-100.

2.4.1 Cationic Surfactant-based Sensors

In electrochemical and biosensor applications, surfactants are mostly used for the modification of the electrode surface. The surfactant molecules adsorb onto the surface of the electrode and offer their respective charges, and for example, cationic surfactants provide positive charges. CTAB is one such cationic surfactant, which was largely used for electrochemical sensors studies. The charged film on the surface of the electrode impacts the rate of charge transfer and also modifies the redox potential of electroactive molecules. Many researchers have utilized cationic surfactant-modified electrodes for the determination of organic molecules and biological species. Several studies have described CTAB-based sensors for the electrochemical determination studies of electroactive species such as caffeic acid, bisphenol-F, dopamine, ascorbic acid, methomyl, uric acid, nilutamide, and hydrogen peroxide.[33,70–75] In the determination of bisphenol A, a hierarchical-metal–organic frameworked glassy carbon electrode (GCE) modified with CTAB surfactant was used and a lower detection limit was found.[76] One of the studies, for the determination of terabromopisphenol A (TBBPA), used long-chain dioctadecyl dimethyl ammonium bromide (DODMA), octadecyltrimethyl ammonium bromide (OTAB), and tri octadecylmethylammonium bromide (TOMA) surfactants for the modification of GCEs. These surfactants activated the sensor by increasing the roughness of the GCE surface. The number of C18 alkyl chains comprising surfactants affected the oxidation of TBBPA differently, and they concluded that the enhancement in the TBBPAs oxidation signal is due to the surfactant's molecular structure, particularly DODMA with two C18 chains, and the limit of detection (LOD) of TBBPA was found to be 1.05 nM.[77] For the determination of morphine (MO), Abraham *et al.* utilized poly(cetyltrimethyl ammonium bromide) and graphene oxide (polyCTAB/GO) for the modification of GCE. The surface area of a polyCTAB/GO-modified GCE is enhanced compared to a poly(cetyltrimethylammonium bromide)-modified GCE. The electrostatic interaction among the tertiary amine group of MO, the cationic group of poly(CTAB), and the nitrogen atom pair delivered and improved electrooxidation of the tertiary amine group of morphine alone.[78] Vinoda *et al.* presented electrochemical studies of statins, *i.e.* atorvastatin (ATRV) by fabricating graphene and CTAB-based electrochemical sensors. In this work the electrochemical oxidation mechanism of ATRV was predicted and the number of electrons and protons involved in the reaction was also studied.[79] Hemn *et al.* prepared a tremendous electrochemical sensor for the determination of rifabutin (RFB) using a boron-doped diamond electrode modified with CTAB and adopted a square-wave voltammetry electrochemical technique. In this work, real sample analysis in the pharmaceutical formulation was done using a CTAB-modified boron-doped diamond electrode.[80]

Kumar *et al.* prepared a modified carbon paste electrode utilizing the cationic surfactant CTAB and it was used for the detection of guanine (GU), UA, and DA simultaneously. The surfactant molecules charged the electrode surface, which influenced the charge transfer rate, altering the oxidation and reduction potentials for electrochemically active molecules. Here, frontier molecular orbital (FMO) and analytical Fukui function concepts were used to identify redox electron transfer (ET) reactive sites on the CTAB, and these theoretical calculations were employed to understand the mechanism of CTAB for the sensitive detection of GU, UA, and DA on the electrode surface. According to their studies, the surfactant (CTAB) on the electrode surface showed very good sensitivity compared to the unmodified electrode. Generally, N^+ in CTAB molecules makes the analytes oxidize rapidly on the surface of the electrode. This concept also supports their mentioned theory based on analytical Fukui functions and FMO. These models are dependable for identifying electron transfer sites of CTA^+. According to this, the trimethylammonium cation (strong electron receiving site) in the surfactant results in the oxidation of a large number of analytes on the electrode surface and aliphatic carbon chain (electron-donating sites) within the surfactant results in the reduction of analytes on the surface (Figure 2.9).

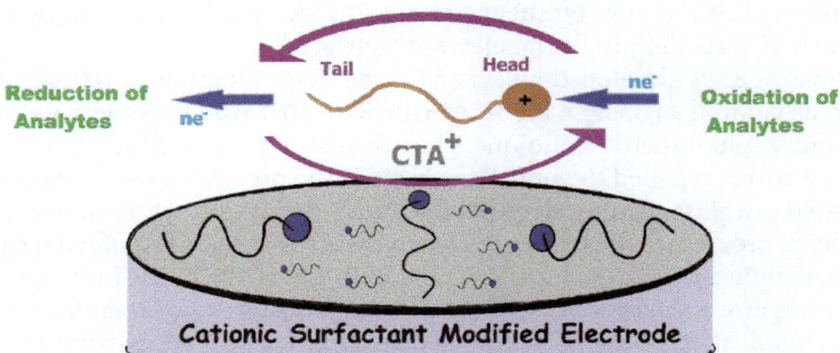

Figure 2.9 Representation of redox-active electron transfer sites on a cationic surfactant that influence the electrochemical oxidation reaction of analytes on the electrode surface.[31]

2.4.2 Anionic Surfactant-based Sensors

It is well known that surfactants have been commonly used in electroanalytical techniques, in that SDS is one of the anionic surfactants frequently utilized in electrochemical studies. For most analytes, it shows good sensitivity when used to modify the electrode surface. Manjunatha *et al.* prepared a multiwall carbon nanotube (MWCNT) paste electrode with SDS by an electrochemical oxidation procedure for the determination of DA, AA, and UA, and studied electrochemical parameters using cyclic voltammetry (CV) and differential pulse voltammetry (DPV) methods. Moreover, the prepared sensor was applied for the determination of DA in a real sample.[81] Similarly, a sodium lauryl sulfate (SLS)-modified carbon nanotube paste electrode (SLSMCNTPE) was prepared for the determination of riboflavin in a pH 7.0 phosphate buffer solution by cyclic voltammetry (CV), resulting in good enhancement of the electrochemical response as compared to a bare carbon nanotube paste electrode. The effects of the concentration of surfactant on the electrode and electrochemical parameters such as scan rate, pH, *etc.* were studied, and the sensor showed good stability, repeatability, reproducibility, and high electrochemical sensitivity.[39] An electrochemical sensor was developed for the determination of an essential anticancer agent cisplatin using MWCNT with COOH-functionalized SDS on a screen-printed electrode and adopted in an anodic stripping voltammetry technique. In the electrochemical analysis the sensor showed good sensitivity.[82] In one of the studies, SDS was added to a solution containing electrolyte with rabeprazole sodium (RAB sodium), an anti-ulcer drug, and the influence of SDS on the voltammetric response of the pencil graphite electrode (PGE) examined. This work concluded that the oxidation peak current enhances at a concentration of 45 μM SDS in the solution. A CV method was used for the various concentration studies of RAB sodium, the LOD being found to be 0.18 μM in the range 0.5–250 μM. At pH 6.0 there is a formation of a cation *via* the protonation of (N) in benzimidazole and RAB sodium pyridine rings, after the adsorption of SDS occurs on the electrode surface, which results in the easy approach of RAB sodium to the electrode surface.[83]

Zabcikova *et al.* developed a sensor for the determination of retinol (also known as vitamin A) using a modified CPE with 30% (w/w) SDS with a differential pulse voltammetric technique. The investigation showed that the surface of the electrode repelled the organic solvent in the presence of surfactants, as compared to a glassy carbon electrode (GCE). CPE-modified SDS demonstrated significant progress in the retinol electroanalysis.[84] Based on the stated results, it was concluded that surfactants could be utilized in CPE for the investigation of pure organic solvents. However, in the same article, they mentioned the electrochemical response of retinyl acetate, retinol and retinylpalmitate in pure aqueous (0.1 M acetate buffer, pH = 4.5) medium, non-aqueous, and aqueous-organic (50% MeCN containing 0.1 M $LiClO_4$) by a CV technique. Because of the lipophilic character of mentioned retinoids, the measurements were carried out in the aqueous medium in an adsorbed state.[84,85]

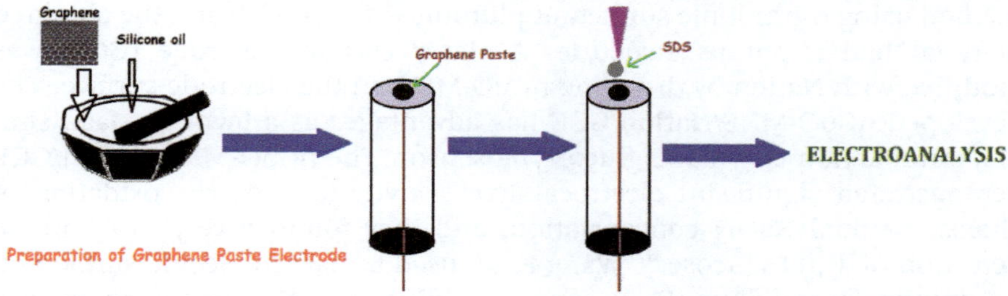

Figure 2.10 Preparation of sodium dodecyl sulfate-immobilized modified graphene paste electrode.

In one of the research works, the author fabricated a graphene paste electrode (GPE) using SDS as a modifier for the sensitive determination of resorcinol (RS) and catechol (CC). The modified GPE was prepared by grinding silicon oil and graphene in an agate mortar. The obtained graphene paste was used to make GPE and then SDS was immobilized on the electrode surface (Figure 2.10). The GPE was compared with other carbon electrodes and it was reported that the adopted methods can be used to easily and rapidly prepare the electrode, with the addition of several substances for electrode modification. The prepared electrode showed a low background current. More significantly, the prepared electrode split the voltammogram peaks for RS and CC which is more difficult for the bare electrode and other modified electrodes. This result could be because of the large surface area and porous structure of SDS/GPE.[29]

2.4.3 Non-ionic Surfactant-based Sensors

In addition to cationic and anionic surfactants, another category of surfactant, *i.e.* non-ionic surfactants is also utilized for analysis by electrochemical methods. Non-ionic surfactants are those that do not contain any charged head group. Examples of non-ionic surfactants are Triton X-100, saponin, polyvinylpyrrolidone, *etc.* Generally, Triton-X 100 (Tx-100) is used as a non-ionic surfactant for the electrochemical determination of several analytes. Stozhko *et al.* utilized Tx-100 surfactant for the modification of a film carbon veil-based electrode (FCVE). The prepared modified electrode Tx-100/FCVE was used for the determination of nitrite. The oxidation peak current of nitrite was enhanced three times compared to unmodified FCVE. The wettability of the carbon screen is inadequate as it has a hydrophobic nature. The hydrophilicity of the surface of the electrode was enhanced by the modification of the electrode with a surfactant; as a result, the surface area increased. They obtained an LOD of 0.01 M with a linear range of 0.1–100 M.[86] Moreover, surfactants have also been utilized as a material for the synthesis of electrode modifiers. Haghparas *et al.* developed a biosensor for the determination of glucose with nano-copper oxide micro hollow spheres (nCuO-MHS), and this material was synthesized by a hydrothermal method using the anionic surfactant pluronic F-127 (F-127) and the obtained material had a porous structure. A glassy carbon electrode (GCE) was modified with Nafion by dropping nCuO-MHS on the electrode surface. The developed nCuO-MHS/Nafion/GCE has advantages as a low cost, fast, simple, durable non-enzymatic glucose biosensor. The nCuO-MHS/Nafion/GCE demonstrated significant electrocatalytic activity towards the oxidation of glucose at dual NaOH concentration, and they found a very low limit of detection of 1 μM glucose.[87] Wang *et al.* used a macromolecule surfactant, *i.e.* polyvinylpyrrolidone (PVP), for the modification of a carbon paste electrode and studied the electrochemical behavior of acyclovir (generally used guanine analog as an antiviral drug). In their work the PVP modified carbon paste electrode enhanced the electrochemical oxidation peak and the method possessed a short analysis time, high sensitivity, handling convenience, and good accuracy. Further, the prepared modified electrode was used practically for real sample analysis.[88]

2.4.4 Zwitterionic and Gemini Surfactant-based Sensors

Zwitterionic surfactants, also called amphoteric surfactants, have both cationic and anionic properties. Based on the structure of amphoteric surfactants, the anionic part is commonly carboxylic, phosphoric or sulfuric acids, the cationic part is generally the quaternary ammonium cations or amines. These molecules display cationic or anionic characters depending on pH.[89] While the anionic character shows in base solutions, cationic properties predominate in acidic conditions. Zapp *et al.* prepared a bio-inspired complex, which was combined with a zwitterionic (ImS3-14) surfactant stabilizing preformed Pd nanoparticles, and the prepared material was coated on a GCE. The prepared material was used as a modifier for the simultaneous determination of dihydroxybenzenes (catechol and hydroquinone), and the bio-inspired surfactant film influenced the catalyzed redox reaction of analyte dihydroxybenzenes. The electrochemical quantification of dihydroxybenzenes was analyzed in cigarette residue as a real sample analysis using square-wave voltammetry. They optimized the electrochemical analysis by means of best result conditions using pH 7.0 phosphate buffer solution.[90] In addition, 3-(*N*,*N*-dimethylpalmitylammonio)propane sulfonate (SB3-16), a zwitterionic surfactant was utilized for the modification of electrode surfaces and applied to an acetylene black paste electrode (ABPE) for the determination of tetrabromobisphenol A (TBBPA). The 0.05 M SB3-16 and TBBPA showed a hydrophobic interaction that enhanced the current response. Especially, the negatively and positively charged zwitterionic surfactant had the capability to become an electronic conductor, thereby facilitating the electron transfer of TBBPA.[91]

Gemini surfactants, also known as dimeric surfactants, are an innovative class of surfactant that consists of a couple of hydrophobic tails linked by a spacer group. These accelerate the electron transfer rate and enhance the sensitivity. As a result, this type of surfactant is the right candidate for electrode modifications. Thus, gemini surfactants show significant potential for usage in gene delivery, inorganic and organic reactions, the pharmaceutical industry, and so on.[92,93] Mo *et al.* made an electrode by modifying butyl-α,ω-bis(dimethyldodecylammonium bromide) gemini surfactant/ionic liquid/MWCNT for the determination of Sudan I in ketchup and hot chili powder samples. The electrode modification displayed satisfactory recovery, beside good repeatability and stability.[93] Moreover, the modifier gemini surfactants endorsed electron transfer and illustrated numerous benefits over conventional surfactants because of their unique properties, such as greater effectiveness in lowering the surface tension of water, lower critical micelle concentration, surface wetting ability, and stronger solubilization.

2.5 Conclusion

Electrochemical techniques are significantly applied because of simple preparation, low cost, high speed, and very good analytical performance. Surfactants are surface-active agents, and a specific class of surfactants is broadly used in electrochemical sensors to develop the electrode and analyte interface properties. Immobilizing surfactants on the surface of electrodes or adding surfactants into electrolytes may significantly improve the selectivity as well as the sensitivity by enhancing the electron transfer rate, changing the charge transfer coefficient and redox potential. In this chapter, we have also presented the types of surfactants, and their structures and roles in the modification of electrodes for the determination of biomolecules, drugs, hazardous materials, *etc.* Surfactant-based electrochemical sensors can be utilized for real sample analysis in different areas, such as food, biomedical, industries and agriculture. We conclude that there has been much research done on surfactant-based electrochemical sensors; however, there are still various limitations and tasks to be overcome. Further research work is required to develop more selectivity in these electrochemical sensors to analyze real samples without the requirement for complex sample preparation methods.

References

1. D. N. Unal, S. Yıldırım, S. Kurbanoglu and B. Uslu, *Trends Anal. Chem.*, 2021, **144**, 116418.
2. J. Fan, L. Ding and Y. Fang, *Langmuir*, 2018, **35**, 326.
3. M. R. Housaindokht, F. J. Fard and N. Ashraf, *J. Surfactants Deterg.*, 2021, **1**, 23.
4. F. Menger and C. Littau, *J. Am. Chem. Soc.*, 1993, **115**, 10083.
5. M. S. Kamal, *J. Surfactants Deterg.*, 2016, **19**, 223.
6. H. Cortés, H. Hernández-Parra, S. A. Bernal-Chávez, M. L. D. Prado-Audelo, I. H. Caballero-Florán, F. V. Borbolla-Jiménez, M. González-Torres, J. J. Magaña and G. Leyva-Gómez, *Materials*, 2021, **14**, 3197.

7. M. Kumar, B. E. K. Swamy, S. Reddy, J. K. S. Kumara and W. Zhao, *J. Electrochem. Soc.*, 2020, **167**, 087511.
8. A. Verdian-Doghaei, M. R. Housaindokht and K. Abnous, *Anal. Biochem.*, 2014, **466**, 72.
9. N. Dey and S. Bhattacharya, *Chem. Rec.*, 2016, **16**, 1934.
10. M. Khavani, M. Izadyar and M. R. Housaindokht, *Sens. Actuators, B*, 2019, **300**, 126947.
11. H. Y. Bi, A. C. Fernandes, S. Cardoso and P. Freitas, *Sens. Actuators, B*, 2016, **224**, 668.
12. M. Kumar, M. Wang, B. E. K. Swamy, M. Praveen and W. Zhao, *Biointerfaces*, 2020, **196**, 111299.
13. S. Reddy, B. E. K. Swamy, S. Aruna, M. Kumar, R. Shashanka and H. Jayadevappa, *Chem. Sens.*, 2012, **7**, 1.
14. T. V. Sathisha, B. E. K. Swamy, K. R. Mahanthesha, M. Kumar and B. Eswarappa, *Chem. Sens.*, 2014, **25**, 1.
15. F. Nasiri, G. H. Rounaghi, N. Ashraf and B. Deiminiat, *Int. J. Environ. Anal. Chem.*, 2021, **101**, 153.
16. G. Ziyatdinova, E. Yakupova, E. Ziganshina and H. Budnikov, *Electroanalysis*, 2019, **31**, 2130.
17. T. V. Sathisha, B. E. K. Swamy, C. C. Vishwanatha, M. Kumar, T. S. Anvekar and B. Eswarappa, *Chem. Sens.*, 2014, **4**(13), 1.
18. J. G. Manjunatha, M. Deraman, N. H. Basri and I. A. Talib, *Arabian J. Chem.*, 2018, **11**(2), 149.
19. P. K. Brahman, R. A. Dar, S. Tiwari and K. S. Pitre, *Colloids Surf., A*, 2012, **396**, 8.
20. G. Ziyatdinova, E. Yakupova, E. Ziganshina and H. Budnikov, *Electroanalysis*, 2019, **31**, 2130.
21. C. Raril, J. G. Manjunatha, G. Tigari and N. Hareesha, *Open Access J. Chem.*, 2018, **2**, 21.
22. M. Sajid, N. Baigand and K. Alhooshani, *Trends Anal. Chem.*, 2019, **118**, 368.
23. V. Rao and Y. Reddy, *Trends Anal. Chem.*, 2017, **9**, 874.
24. M. Kumar, B. E. K. Swamy, T. V. Sathisha and R. Deepa, *Anal. Bioanal. Electrochem.*, 2015, **7**(5), 610.
25. A. F. Khan, D. A. C. Brownson, E. P. Randviir, G. C. Smith and C. E. Banks, *Anal. Chem.*, 2016, **88**, 9729.
26. E. Nagles, O. García-Beltr and J. A. Calderon, *Electrochim. Acta*, 2017, **258**, 512.
27. M. Kumar and B. E. K. Swamy, *Mater. Sci. Eng.*, 2016, **C58**, 142.
28. T. S. S. K. Naik, B. E. K. Swamy, C. C. Vishwanath and M. Kumar, *J. Anal. Bioanal. Tech.*, 2015, **6**(6), 1.
29. J. G. Manjunatha, *Chem. Data Collect.*, 2020, **25**, 100331.
30. V. Erady, R. J. Mascarenhas, A. K. Satpati, A. K. Bhakta, Z. Mekhalif, J. Delhalle and A. Dhason, *Microchem. J.*, 2019, **146**, 73.
31. M. Kumar, Y. Fu, M. Wang, B. E. K. Swamy, G. K. Jayaprakash and W. Zhao, *J. Mol. Liq.*, 2021, **321**, 114893.

32. N. O. Decarli, E. Zapp, B. S. de Souza, E. R. Santana, J. P. Winiarski and I. C. Vieira, *Biochem. Eng. J.*, 2022, **186**, 108565.
33. S. S. Shankar, B. E. K. Swamy, U. Chandra, J. G. Manjunatha and B. S. Sherigara, *Int. J. Electrochem. Sci.*, 2009, **4**, 592.
34. R. Hosseinzadeh, R. E. Sabziand and K. Ghasemlu, *Biointerfaces*, 2009, **68**, 213.
35. M. Kumar, B. E. K. Swamy, S. Reddy, T. V. Sathisha and J. Manjanna, *Anal. Methods*, 2013, **5**, 735.
36. P. R. Vernekar, B. Purohit, N. P. Shetti and P. Chandra, *Microchem. J.*, 2021, **160**, 105599.
37. M. Kumar, B. E. K. Swamy, U. Chandra and A. W. Gebisa, *Int. J. Nanotechnol.*, 2017, **14**, 930.
38. A. E. Waleed, D. Hajjar, A. A. Makki, E. A. Elshehy and J. W. Choi, *Int. J. Environ. Anal. Chem.*, 2021, **101**, 1.
39. G. Tigari, J. G. Manjunatha, C. Raril and N. Hareesha, *ChemistrySelect*, 2019, **4**, 2168.
40. M. Kumar, B. E. K. Swamy, M. H. M. Asif and C. C. Viswanath, *Appl. Surf. Sci.*, 2017, **399**, 411.
41. A. Zahid, A. Lashin, U. A. Rana, N. Al-Arifi, I. Ullah, D. D. Dionysiou, R. Qureshi, A. Waseem, H. B. Kraatz and A. Shah, *Electrochim. Acta*, 2016, **190**, 1007.
42. U. Chandra, B. E. K. Swamy, M. Kumar and A. W. Gebisa, *Int. J. Nanotechnol.*, 2017, **14**, 739.
43. M. Kumar, B. E. K. Swamy, S. Reddy, W. Zhao, S. Chethana and V. G. Kumar, *J. Electroanal. Chem.*, 2019, **835**, 96.
44. J. S. Stefano, L. O. Orzari, H. A. Silva-Neto, V. N. de Ataíde, L. F. Mendes, W. K. T. Coltro, T. R. L. C. Paixão and B. C. Janegitz, *Curr. Opin. Electrochem.*, 2022, **32**, 100893.
45. S. Nambudumada, J. Prinith, G. Manjunatha, M. D. Albaqami, A. M. Tighezza and M. Sillanpää, *ChemistrySelect*, 2022, **7**, 202203572.
46. G. K. Ziyatdinova, E. R. Ziganshina and H. C. Budnikov, *J. Anal. Chem.*, 2012, **67**, 869.
47. C. B. B. Farias, F. C. G. Almeida, I. A. Silva, T. C. Souza, H. M. Meira, R. C. F. S. Silva, J. M. Luna, V. A. Santos, A. Converti, I. M. Banat and L. A. Sarubbo, *Electron. J. Biotechnol.*, 2021, **51**, 28.
48. K. Schumann and K. Siekmann, Soaps, in *Ullmann's Encyclopedia of Industrial Chemistry*, Wiley-VCH, Weinheim, 2005.
49. D. F. H. Wallach, R. Mathur, G. J. M. Redziniak and J. F. Tranchant, *J. Soc. Cosmet. Chem.*, 1992, **43**, 113.
50. Á. G. García, E. M. Timmers, N. Romijn, S. Song, S. Sahebali, R. Tuinier and I. K. Voets, *Colloids Surf., A*, 2019, **561**, 201.
51. K. Weber and D. J. Kuter, *J. Biol. Chem.*, 1971, **246**, 4504.
52. J. Cross and E. J. Singer, *Cationic Surfactants: Analytical and Biological Evaluation*, CRC Press, Florida, 1st edn, 1994.
53. E. Busseron, Y. Ru, E. Moulin and N. Giuseppone, *Nanoscale*, 2013, **5**, 7098.

54. L. Y. Zakharova, T. N. Pashirova, S. Doktorovova, A. R. Fernandes, E. Sanchez-Lopez, A. M. Silva, S. B. Souto and E. B. Souto, *Int. J. Mol. Sci.*, 2019, **20**, 5534.
55. D. G. Hayes and G. A. Smith, *Biobased Surfactants*, AOCS Press, Illinois, 2nd edn, 2019, p. 3.
56. B. E. Rapp, *Micro and Nano Technologies, Microfluidics: Modelling, Mechanics and Mathematics*, Elsevier, Amsterdam 2017, p. 421.
57. J. G. Manjunatha, B. E. K. Swamy, O. Gilbert, G. P. Mamatha and B. S. Sherigara, *Int. J. Electrochem. Sci.*, 2010, **5**, 682.
58. M. A. Moselhy, E. G. Zaki, S. A. E. H. El-Maksoud and M. A. Migahed, *ACS Omega*, 2022, **7–36**, 32014.
59. S. Tamjidi, B. K. Moghadas, H. Esmaeili, F. ShakerianKhoo, G. Gholami and M. Ghasemi, *Process Saf. Environ. Prot.*, 2021, **148**, 775.
60. V. B. Fainerman, D. Möbius and R. Miller, *Studies in Interface Science*, Elsevier Publishing, Amsterdam, 1st edn, 2001.
61. M. Ruiz-Peña, T. P. Oropesa-Nuñez, S. R. W. Louro and A. Pérez-Gramatges, *Colloids Surf., B*, 2010, **75**, 282.
62. S. Morandat and K. El Kirat, *Colloids Surf., B*, 2007, **55**(2), 179.
63. M. S. Kamal, *J. Surfactants Deterg.*, 2016, **19**, 223.
64. S. De, S. Malik, A. Ghosh, R. Saha and B. Saha, *RSC Adv.*, 2015, **5**, 65757.
65. I. M. Banat, Q. Carboue, G. Saucedo-Castaneda and J. J. Cazares-Marinero, *Bioresour. Technol.*, 2021, **320**, 124222.
66. C. B. B. Farias, F. C. G. Almeida, I. A. Silva, T. C. Souza, H. M. Meira, R. C. F. S. Silva, J. M. Luna, V. A. Santos, A. Converti, I. M. Banat and L. A. Sarubbo, *Electron. J. Biotechnol.*, 2021, **51**, 28.
67. N. S. Prinith, J. G. Manjunatha and N. J. Hareesha, *J. Iran. Chem. Soc.*, 2021, **18**, 3493.
68. S. S. Shankar, B. E. K. Swamy, K. R. Mahanthesha, C. C. Vishwanatha and M. Kumar, *Anal. Bioanal. Electrochem.*, 2013, **5**(5), 555.
69. J. G. Manjunatha, *Heliyon*, 2018, **4–11**, 986.
70. V. Erady, R. J. Mascarenhas, A. K. Satpati, A. K. Bhakta, Z. Mekhalif, J. Delhalle and A. Dhason, *Microchem. J.*, 2019, **146**, 73.
71. J. G. Manjunatha, B. E. K. Swamy, R. Deepa, V. Krishna, G. P. Mamatha, U. Chandra, S. S. Shankar and B. S. Sherigara, *Int. J. Electrochem. Sci.*, 2009, **4**(5), 662.
72. G. Manasa, R. J. Mascarenhas, A. K. Satpati, B. M. Basavaraja and S. Kumar, *Colloids Surf., B*, 2018, **170**, 144.
73. C. C. G. Silva and D. D. Souza, *Talanta*, 2018, **189**, 389.
74. S. Yang, S. Ding, L. Li, S. Ding, Q. Cao, J. Yang, W. Xu and A. Chen, *J. Electrochem. Soc.*, 2017, **164**, B710.
75. Y. Temerk, H. Ibrahim and N. Farhan, *Anal. Methods*, 2015, **7**, 9137.
76. J. Zhang, X. Xu and L. Chen, *Sens. Actuators, B*, 2018, **261**, 425.
77. X. Chen, L. Ji, Y. Zhou, S. Xia, J. Tong and K. Wu, *Electrochim. Acta*, 2016, **190**, 490.
78. P. Abraham, S. Renjini, T. E. M. Nancy and V. A. Kumary, *J. Appl. Electrochem.*, 2020, **50**, 41.

79. V. B. Patil, S. J. Malode, S. M. Tuwar and N. P. Shetti, *Sens. Int.*, 2022, **3**, 100198.
80. H. A. H. Barzani and Y. Yardım, *Diamond Relat. Mater.*, 2023, **132**, 109658.
81. J. G. Manjunatha, M. Deraman, N. H. Basri, N. S. M. Nor, I. A. Taliband and N. Ataollahi, *C. R. Chim.*, 2014, **17**, 465.
82. E. M. Materon, A. Wong, S. I. Klein, J. Liu and M. D. P. T. Sotomayor, *Electrochim. Acta*, 2015, **158**, 271.
83. P. Y. Khashaba, H. R. Ali and M. El-Wekil, *Anal. Bioanal. Electrochem.*, 2017, **9**, 729.
84. S. Žabčíkova, T. Mikysek, L. Červenka and M. Sýs, *Food Technol. Biotechnol.*, 2018, **56**, 337.
85. M. Sýs, S. Žabčíková, L. Červenkaand and K. Vytřas, *Potravinarstvo Slovak J. Food Sci.*, 2016, **10**(1), 260.
86. N. Y. Stozhko, M. A. Bukharinova, E. I. Khamzina, A. V. Tarasov and S. V. Sokolkov, *Chemosensors*, 2020, **8**, 78.
87. Z. Haghparas, Z. Kordrostami, M. Sorouri, M. Rajabzadeh and R. Khalifeh, *Biotechnol. Bioprocess Eng.*, 2020, **25**, 528.
88. P. Wang, T. Gan, J. Zhang, J. Luo and S. Zhang, *J. Mol. Liq.*, 2013, **177**, 129.
89. S. Tamjidi, B. K. Moghadas, H. Esmaeili, F. S. Khoo, G. Gholami and M. Ghasemi, *Process Saf. Environ. Prot.*, 2021, **148**, 775.
90. E. Zapp, F. D. Souza, B. S. Souza, F. Nome, A. Neves and I. C. Vieira, *Analyst*, 2013, **138**, 509.
91. X. Wei, Q. Zhao, W. Wu, T. Zhou, S. Jiang, Y. Tong and Q. Lu, *Sensors*, 2016, **16**, 1539.
92. M. S. Kamal, *J. Surfactants Deterg.*, 2016, **19**, 223.
93. Z. Mo, Y. Zhang, F. Zhao, F. Xiao, G. Guo and B. Zeng, *Food Chem.*, 2010, **121**, 233.

CHAPTER 3

Surfactant Sensors for Bio-based Sensing

PRIYA CHAUHAN,*[a] G. MADHAVI*[b] AND
VEERA MANOHARA REDDY YENUGU*[c]

[a] School of Studies in Environmental Chemistry, Jiwaji University, Gwalior-474011, India; [b] Department of Chemistry, Sri Venkateswara University, Tirupati-517502, India; [c] Department of Chemistry, Sri Venkateswara College, University of Delhi, New Delhi 110021, India
*Emails: priyachauhan0129@gmail.com; gmchem01@gmail.com; yvreddy@svc.ac.in

3.1 Introduction

The world seems to have a natural division between chemical and physical sensors. However, there are those that do not classify easily, like relative humidity sensors, a chemical sensor traditionally lumped in with physical sensors. Also, sensors are often discussed along with the topic of actuators. Chemical sensors have a chemical or molecular target to be measured. Biosensors are defined as sensors that use biomolecules and structures to measure something with biological significance or bioactivity. More appropriately, biosensors target a biomolecule of interest for measurement. The biosensor can usually be considered a subset of chemical sensors because the transduction methods, sometimes referred to as the sensor platforms, are the same as those for chemical sensors. Chemical sensor arrays with instrumentation, having popular names like the electronic nose or electronic tongue, have been constructed to address chemically complex analytes like taste, odor, toxicity, or freshness.[1]

A useful definition for a chemical sensor is "a small device that as the result of a chemical interaction or process between the analyte gas and the

Detection Science Series No. 23
Surfactant-based Sensors in Chemical and Biochemical Detection
Edited by J. G. Manjunatha

Published by the Royal Society of Chemistry, www.rsc.org

sensor device, transforms chemical or biochemical information of a quantitative or qualitative type into an analytically useful signal."

Tensides or surfactants are surface-active agents. There are four main types of surfactant: anionic, cationic, nonionic and amphoteric. They are used for washing, cleaning, and disinfection in a broad variety of industries.[2] Rapid industrialization, population growth, and increased standard of living have caused constant growth in surfactant demands. The growing demand for anionic surfactants is the result of the increased need for home care and personal care applications, and recently, for electric vehicles.[3–5] Nonionic surfactants are amphipathic molecules. They consist of a lipophilic part (long alkyl chain or fatty acid) and a hydrophilic part (ethylene oxide) of varying length. Nonionic surfactants do not ionize in water. Surfactants have been widely used in the preparation of nanoparticles (NPs) because they are electron-rich and provide good chelating centers for improving the stability of NPs synthesized in solution.[6] One of the common surfactants for the synthesis of metallic NPs is a gemini surfactant, which consists of two hydrophobic tails and two hydrophilic heads connected by a spacer.[7,8] Gemini surfactants are characterized by their relatively low critical micelle concentrations and high efficiency in decreasing the surface and interfacial tensions of water compared to conventional monomeric surfactants.[9] These properties provide a high adsorption affinity toward the surface of metal NPs for high stabilization.

In household applications, anionic surfactants are used in detergents or emulsifiers for a variety of home products for washing and cleaning. They are utilized regularly in combination with nonionic surfactants to enhance cleaning properties. They are also employed as wetting agents in oil-in-water systems in beauty products to achieve stable emulsions with the pigments and fillers dissolved in the water phase.[10]

Because of the reduction of the water tension, prevention of gas exchange on the water surface and eutrophication, anionic surfactants have a negative influence on the environment. They also induce the disintegration of cell membranes. Thus, the development of new, simple, reliable, and robust methods to control the levels of anionic surfactants in a variety of products during their production and quality control, as well as in the environment is paramount and essential. The classical method for the determination of low concentrations of anionic surfactants is based on methylene blue active substances (MBASs), whereas high concentrations can be measured by two-phase titration.[11] Both methods are widely used but they have many drawbacks, including their time-consuming character, employment of toxic chemicals, and requirement for a well-trained and experienced analyst. In contrast, ion-selective surfactant sensors could overcome the above-mentioned drawbacks, as they are simple, low-cost, reliable, and fast.

However, the most of practical methods for the determination of diverse analytes, including anionic surfactants (ASs) are based on the employment of ion-selective electrodes (ISEs). Generally, sensors for ASs must be fast, reliable, and cost-effective for routine analysis[12] and ISEs meet almost all of these requirements.[13] ASs are widely used as the main constituents in detergent formulations. Consequently, AS waste is one of the most widespread aquatic pollutants, which imposes a severe requisite to develop inexpensive, accurate, and practical methods for detecting ASs.[14,15] From this point of view, the use of potentiometric sensors seems to be a highly promising method.

Except for solid-state surfactant sensors,[16–18] they are usually based on a PVC liquid-type sensor membrane incorporated into the electrode body or coated on a wire.[19] The sensor membrane usually consists of a high molecular weight PVC and a plasticizer in a weight ratio of 1 : 2, and an ionophore. The plasticizer softens the rigid PVC and allows the ionophore to be dissolved in a highly lipophilic medium. The ionophore is an ion-pair synthesized by the reaction of high molecular weight anionic and cationic ions and/or surfactants. In order to prevent leaching in water and to decrease the lifetime, the ionophore should have very low solubility in water and high lipophilicity, whereas to increase the sensitivity and the lifetime, high stability is needed, leading to a positive influence on sensor properties such as higher sensitivity and broader effective concentration range.

The development of the new ionophores is of crucial importance for the advancement of novel surfactant sensors with better properties. Recently, our group and others showed that an increase in the active surface and electrotransfer may be achieved by the employment of nanomaterials as ionophores that also provide higher signal stability and lifetime. As nonionic surfactants have a negative influence on potentiometric titrations and end-point detection,[20,21] it is important to investigate the influence of nonionic surfactants on potentiometric curve shape, size, and especially, the detection of the end-point.

The composition of ISEs for the determination of surfactant concentration is based on three components: an ion-pair associate as a sensor material (two opposite charge surfactants), PVC, and plasticizer. Consequently, sensitivity, selectivity, and lifetime of the ISEs will be mainly defined by the properties of the ionophore. As the lifetime of ISEs is limited and mainly depends on the water-solubility of membrane components, the development of new, highly lipophilic ionophores is a promising approach to ensure the best features of potentiometric sensors. Also, new selective ionophores can be applied in transduction principles other than potentiometry, such as voltammetry.[22]

For the construction of ISEs for AS detection, different ion-pair associates have been employed: tetrahexadecylammonium-dodecyl sulfate,[23] 1,3-didecyl-2-methyl-imidazolium-tetraphenylborate,[24] hexadecyltrioctadecylammonium-tetraphenylborate,[25] cetyltrimethylammonium-dodecylsulfate,[26] dodecyltrimethylammonium-dodecylbenzenesulfonate,[27] cetyltrimethylammonium-tetraphenyl borate[28] hyamine-tetraphenylborate,[29] and others. Cationic surfactants in the form of quaternary alkylammonium compounds (QAC) are used as a counter-ions responsible for interaction with the target analytes (ASs). As the nitrogen atoms are cationic, and the negative charge is usually located on the halogens or other anions, QAC molecules possess large dipole moments. The charged quaternary nitrogen atoms of QACs can be also within aromatic or aliphatic rings.[30] The hydrophobic part is ordinarily a longer alkyl chain, while the hydrophilic group is defined by the surfactant type. In addition, QACs are often used as biocides[31] and low-level disinfectants in healthcare facilities,[32] as they can exhibit perceptible antimicrobial activities.

Due to these properties, QACs are widely applied as disinfectants and for water treatment to control macro-fouling of water systems. In recent years the number of applications of QACs has considerably increased. They are also used as phase-transfer catalysts,[33] green solvents based on ionic liquids,[34] and electrolytes for liquid batteries.[35] One of the major issues in the construction of ISEs is reliable immobilization of the electroactive compounds (EAC) in the PVC matrix. Leaching of the EAC causes severe changes in the sensor response and shortens the lifetime of the sensor.[36] This problem can be solved by modification of the membrane composition and the employment of a less water-soluble, highly lipophilic EAC.[37] The reduced solubility of these compounds may be accomplished by alkylation with considerably long alkyl groups that will also suppress the leaching of EAC from the membrane and, thus, positively affect the response stability and the lifetime of the sensor.

The ionophore's capability to act as a sensor is mostly governed by potential and capacity to accomplish complex chemical interactions with analytes in their binding sites. The properties of an ionophore are determined by a sensitive balance between rigidity, often required for the ion-selectivity, and flexibility necessary for fast ion transfer and electrode response. These properties can be predicted by QSAR models and numerically expressed as the average molecular polarizability (α).[38] Indeed, the strength of the interaction between the ionophore and the ionic analyte is related to polarizability[39] and, thus, it can be used to evaluate the sensory properties. As potentiometric sensors for AS determination are frequently constrained to several ionophores for selective recognition, particular concern must be given to discovery of new ionophores with improved properties.

Moreover, new quaternary alkylammonium compounds have been synthesized and tested as sensing elements for surfactant selective electrodes. Sensitive potentiometric sensors for ASs were prepared based on the new QAC-tetraphenylborate (TPB) ion-pairs. To test the performance of the new sensing materials, five electrodes with a liquid-type membrane were prepared. The presented research aimed to increase the hydrophobicity of the sensing materials and improve their immobilization in order to obtain more stable membrane compositions and better analytical characteristics. The applicability of the new QACs as sensor materials was demonstrated *via* AS determination in aquatic solutions and real systems.[40]

Recently 1,3-dihexadecyl-1*H*-benzo[*d*]imidazol-3-ium–tetraphenylborate (DHBI–TPB) was synthesized and used as a new long-chain quaternary ammonium cation with high amphipathic character to detect cationic surfactants in personal care products and disinfectants.[41] The aim of this research was to investigate the performance of the developed DHBI–TPB surfactant sensor for the detection and quantification of anionic surfactants. Studies on the influence of nonionic surfactants on the device performance were also performed. Finally, newly developed DHBI–TPB electrodes were employed for the quantification of anionic surfactants in detergents and household care products.

3.2 Interactions of Anionic Surfactant Systems

Although ISEs are most commonly used for analysis of ASs in the premicellar region (monomeric surfactant) of surfactant solutions, they can also be used as probes for thermodynamic studies of surfactant solutions. Thus, ISEs can be employed to investigate the dimerization, aggregation, and micelle formation of ASs in solution. The binding isotherm of ASs bound to other materials can also be calculated.[42] When membrane electrodes are used for direct potentiometry, rather than as titration end-point sensors, the polymer matrix and solvent can modify the thermodynamic properties of the ASs that are measured; in particular, the critical micelle concentration (CMC) can be reduced significantly. This has implications for any studies of, *e.g.* post CMC monomer activity. Cationic/anionic cat-anionic surfactant mixtures are typical.[43] However, there are no published results in which ISEs are employed to study the interactions of cat-anionic mixtures. ISEs are usually used for characterization of ionic/nonionic surfactant systems.[44] Interaction studies are mainly focused on dodecyl sulfate (DS).[45,46] Industrial formulations often contain mixed surfactant/polymer systems with superior characteristics (enhanced surface activity, improved rheological and interfacial properties) to those of the pure component systems. Superior characteristics of this blended systems are utilized in the creation of new formulations of detergents and pharmaceutical products.[47] Therefore, considerable efforts have been made to characterize supramolecular aggregation of surfactants, and polymers, and their mixed systems. Adsorption of ASs on water-soluble polymers: neutral polymers,[48] polyelectrolytes,[49] biopolymers,[50] and biomolecules can also be monitored using ISEs (surfactant-binding isotherms). Effects of electrostatic factors, polymer molar mass, surfactant counter-ions, temperature, pH, and structural changes are considered. Potentiometric measurements based on the use of ion-selective electrodes have been also used to study the adsorption of ASs with functionalized single-walled carbon nanotubes (SWCNTs).[51] This interaction is driven by hydrophobic forces between the hydrocarbon tails of the ASs and the nanotube walls. An interesting approach for indirect determination of surfactants using enzymatic activity was published by Khalaed *et al.*[52] They described a simple potentiometric enzymatic method for analysis of surfactants based on their inhibitory effect on acetylcholinesterase enzyme activity. Hydrolysis of acetylcholine was measured using a disposable acetylcholine potentiometric sensor. Various cationic, anionic, and nonionic surfactants were determined in the concentration range 0 to 40 $mg\,mL^{-1}$, with LODs reaching 0.07 $mg\,mL^{-1}$. This method was used for the determination of surfactants in pharmaceutical formulations, commercial detergents, and environmental samples.

3.3 Application of Surfactants in Voltammetric Methods

Surfactants have been widely used for various purposes in electrochemistry and electroanalysis. In voltammetric measurements, surfactants as modifiers have been reported to enhance the selectivity and sensitivity of the process. Surfactants can be added either into the electrolyte solution or to the electrode composition. Often, the experimental method for modifying carbon-based electrodes can be simplified by adding a surfactant (modifying agent) to the background electrolyte, so-called *in situ* modification. This method has the critical advantage of diminishing or removing the required electrode preparation steps before the analysis.[53]

Organic amphiphilic compounds consisting of a hydrophobic head and a tail are referred to as surfactants. The polar heads of surfactants can be nonionic or ionic, and subsequent classes can be divided into cationic, anionic, and zwitterionic.[54] Nonionic surfactants, usually composed of a polar region, have no charge in the aqueous medium. Positively and negatively charged surfaces are formed by using cationic and anionic surfactants, respectively. Zwitterionic amphoteric surfactants produce a positive or negative charge depending on the pH value of the solution.[55] Surfactants as organized assemblies are significantly applied in ordinary life, the chemical industry, medical science, material processing, and the have considerable scope for use in nano reactors, drug carriers, and the fabrication of sensors.[56–59] A sensor detects and converts a variable value to an electronic signal. The most important criteria for evaluating sensor performance are its sensitivity, selectivity, and stability. Generally, techniques such as spectrophotometry, colorimetric assay,[60] or fluorescence spectrometry[61] are employed to design sensors. However, they may require extensive sample preparation and have difficulty detecting small levels of analytes due to matrix interference and insufficient sensitivity. In addition, they may require professional user training. To overcome the limitations, attention has been paid to their ability to achieve high sensitivity and selectivity using electrochemical methods. Today, electrochemical sensors have become an essential category of chemical sensors, which have many advantages, including suitable field-based size, absolute discrimination ability, and fast response.[62–64] They have received much consideration, both in fundamental research and practical applications, because of several benefits, such as easy construction, reasonable cost, and outstanding detection limits, sufficient to handle complicated biofluids.[65–67] Voltammetry is an important electrochemical technique on which many electrochemical sensors rely. In this case, the selectivity of the sensor is generally perfect, because the analyte can be easily detected with a voltammetric peak potential. Furthermore, voltammetry with modified electrodes has been widely used in voltammetric sensors to ensure high analytical and operational characteristics for the determination of analytes.[68,69] Currently, much attention is paid to the use of surfactants in the construction of modified electrodes. Various outcomes arise when surfactants are adsorbed onto the surface of electrodes. Examples are the higher transfer rate of electrons and changes in the charge transfer coefficients and redox potentials, enhancing their ability to amplify signals.[70–72] Furthermore, surfactants have a wide range of properties that render them a viable option for monitoring the selective features of sensor response to the analytes; this is owing to the ability of surfactants to alter the electrode surface in terms of hydrophilic, lipophilic, and electrostatic features.[73,74] Also, surfactants are used as an anti-fouling agent to overcome the surface fouling problem of some electrochemical sensors. In this regard, cetyltrimethyl ammonium bromide (CTAB) has been utilized in the construction of a carbon paste electrode modified with magnetic molecularly imprinted nanoparticles for the determination of bisphenol A (PBA). The presence of CTAB as an ingredient in the carbon paste composition considerably prevents the fouling of the electrode surface.[75] Additionally, diverse kinds of surfactants have been used as additives to control the size and shape of metallic nanostructures to produce highly ordered nanostructures for electrochemical sensing purposes. For instance, anionic cocamidopropyl betaine (CAPB), nonionic decyl glucoside (DG), and cationic cetyltrimethylammonium chloride (CTAC) surfactants have been applied to the process of anodizing a gold electrode to prepare a nanoporous gold film (NPGF) for the electrochemical sensing of quercetin. CAPB proved to be best for the synthesis of a highly electrocatalytic NPGF.[76]

The unique catalytic, electronic, magnetic, and high surface-to-volume ratio properties of nanoparticles make them beneficial components for modified electrodes due to the acceleration of the electron transfer rate and the lowering of the required overvoltage for the electrochemical process. The most crucial nanomaterial modifiers mainly include carbon nanomaterials and metallic nanoparticles, as well as their oxides, and diverse compounds.[77,78]

Polymeric materials, especially those that are conductive, have been considered by many researchers for constructing modified electrodes. Polymers play a crucial role in the construction of surfactant-based voltammetric sensors. Moreover, these compounds have exhibited positive effects on reducing overvoltage, improving electron transfer, and increasing current.[79] On the other hand, remarkable conductivity properties have been attributed to composite materials, which are associated with the synergistic interconnections of their elements.[80,81] For instance, studies have confirmed that electronic charge flow can be facilitated in a composite using metal nanoparticles and polymer; this feature is of particular interest for electrochemical purposes. So far, various polymers and nanomaterials have been used to construct modified electrodes, creating composite and hybrid electrodes. Also, surfactants ensure the stability of a suspension of nanoparticles and affect the voltammetric characteristics of analytes.[82,83]

Voltammetric techniques are broadly classified into nonstripping and stripping techniques. The first category involves cyclic voltammetry (CV), linear sweep voltammetry (LSV), differential pulse voltammetry (DPV), and square-wave voltammetry (SWV). The second class, that is, stripping voltammetry, encompasses extremely sensitive electrochemical techniques that utilize a preconcentration (accumulation) step to enhance the sensitivity of determination. Stripping techniques include anodic stripping voltammetry, cathodic stripping voltammetry (CSV), and adsorptive stripping voltammetry (AdSV). Stripping voltammetry is preferred over the nonstripping techniques for the sensitive determination of various compounds owing to benefits such as lower sensitivity to matrix effects compared to other electroanalytical methods.[84,85]

3.4 Surfactant Sensors in Electrochemical Systems

Metallic nanoparticles (NPs) are used in various fields of science including optics, catalysis, biochemistry, medicine, and chemical sensors.[86–90] One of the major problems associated with NPs is their tendency to aggregate, leading to a decrease in the surface area per atom and to the loss of interesting size-dependent properties. To avoid NP aggregation different materials have been used, such as polymers, ionic liquids, and surfactants. The use of surfactants is very attractive since it allows the preparation of NPs in water or in reversed micelles, the latter method allowing better control of the particle diameter and shape.[91,92]

Surfactant-stabilized NPs are very attractive materials for the construction of electrochemical sensors. The advantages of these materials are associated with both the metallic core and the stabilizer employed: the former allows fast electron-transfer processes and the latter allows the deposition of a surfactant film on the solid electrode, which results in immobilization of the NPs. Furthermore, it has been shown that the use of surfactants alters the electrode–solution interface and subsequently influences the electrochemical processes of reactive substances.[93] Thus, surfactants have been used successfully to improve the limits of detection of various analytes through a mechanism involving "synergistic adsorption", *i.e.*, the possibility of improving analyte adsorption on the electrode surface due to the presence of a surfactant film, facilitating electron transfer between the analytes and the electrode.[94] Additionally, surfactants allow the immobilization of biomolecules or bio-inspired molecules through the formation of multiple surfactant layers which entrap biomolecules on the electrode surface. This strategy provides excellent stabilization and immobilization of natural or mimetic enzymes, thus providing a longer biosensor lifetime.[95]

Surfactants are well-known amphiphilic molecules and can form various supramolecular assemblies, such as micelles and vesicles in aqueous solutions.[96] These assemblies are heterogeneous and possess a hydrophobic core and a hydrophilic surface. These microheterogeneous structures enable surfactant aggregates to encapsulate lipophilic fluorophores and to construct fluorescent sensors with high sensitivity and selectivity.[97,98] A number of advantages have been reported in using surfactant aggregates for developing fluorescent sensors. For instance, the sensitivity of a fluorescent molecular sensor cannot only be enhanced by using surfactant assemblies, but also be easily tuned by simply adjusting the coexisting surfactant concentration.[99–101] Moreover, surfactant assemblies with surface charges can function as noncovalent receptors to counter-ions. For example, aggregates of an anionic surfactant such as sodium dodecyl sulfate (SDS) have been used in fabricating fluorescent sensors for metal ions such as Cu^{2+} and Hg^{2+}.[102,103] Conversely, assemblies of cationic surfactants like dodecyl trimethyl ammonium bromide (DTAB) have been used in developing fluorescent sensors for anionic ions such as S^{2-}, F^-, and CN^-.[104,105]

In addition, the container properties of surfactant assemblies have been recently used in developing fluorescent sensor arrays with pattern recognition ability. Anslyn and co-workers reported the use of Tween 80 micelles as containers to dissolve four different types of concentrated fluorophores and fabricate a sensor array providing a seven-signal pattern for a typical explosive analyte.[106] Thayumanavan and co-workers developed a strategy for the use of the complexes of surfactants and polyelectrolytes to obtain supramolecular assemblies for encapsulating different fluorophores to provide pattern recognition for proteins. The addition of proteins can disassemble the complexes and release those encapsulated fluorophores, which results in various fluorescence changes for different fluorophores. Moreover, by simply varying the structure and charge of surfactant and polyelectrolyte, they realized the pattern recognition of proteins using a single fluorescent guest.[107] The non-covalent interaction between the fluorophore and surfactant assemblies provides a simple but powerful way to tune the sensory performance.

It is known that many heavy metal ions are highly toxic and are involved in serious diseases. Therefore, significant attention has been given to developing fluorescent sensors and arrays for the detection of multivalent transition metal ions (especially, heavy metal ions) due to increasing concerns about environmental protection and human health.[108–110] However, most of the reported fluorescent sensor arrays for metal ions have focused on using different fluorophores to derive and synthesize multiple sensor elements. As mentioned earlier, this strategy requires tedious synthesis efforts and a complex data-collecting process. It is in great demand, and a big challenge, to develop fluorescent sensors that are capable of pattern recognition of multiple metal ions in aqueous solutions.

3.4.1 Fluorescent Surfactant Sensors

Fluorescent molecular sensors have attracted increasing attention in fundamental research work and practical applications due to their high sensitivity and selectivity for detecting analytes. A typical fluorescent molecular sensor usually consists of a fluorophore for signal output, a receptor to bind analytes, and a spacer to chemically link these two functional units.[111] For detecting various target analytes, different selective receptor units usually need to be specially designed and chemically derived for a selected fluorophore using an organic spacer. Although numerous fluorescent sensors with high sensitivity and selectivity have been developed, it is still a big challenge to meet the requirements of detecting the large variety of analytes of interest in the fields of environmental protection, food and medicine safety, and medical diagnosis. Inspired by the properties of sensing smell and taste, researchers have made efforts towards developing fluorescent sensor arrays containing a series of sensor elements to provide recognition patterns to different analytes.[112] This type of an array has fewer requirements for the selectivity of the receptor of each sensor element, and depends more on the cross-reactive properties of the sensor elements used. Usually, different fluorophores or different receptors are chosen to generate an array of sensors, and to provide cross-reactive responses to analytes.[113] Combining the differential responses from each of these sensor elements to a typical analyte can generate a specific recognition pattern for that analyte, and such a sensor array possesses high throughput in sensing and identifying different analytes. However, although this method reduces dependence on synthetic efforts, better discriminative properties of an array depend on using large numbers of sensor elements. This, to some extent, restricts the applicability of pattern recognition. A suitable, but simple and powerful, technique to tune the sensory capability without the need for heavy synthetic efforts is attractive for developing fluorescent sensors with pattern recognition abilities.[114–116]

3.4.2 Potentiometric Surfactant Sensors

Potentiometric surfactant sensors, ion-selective electrodes (ISEs), have many advantages in practical applications, such as reduced size, higher sensitivity and reliability, no sample destructive nature, and portable instrumentation at a low price. Potentiometric sensors are applicable in diverse experimental conditions, which makes them appealing for applications in environmental monitoring and industrial process control, *etc.*[117,118]

Potentiometric surfactant sensors are ordinarily used as titration endpoint sensors in batch mode, although they can also be used continuously in flow injection analysis (FIA) mode.[119,120] Surfactant selective electrodes also serve as probes for thermodynamic studies to oversee the aggregation processes of surfactants to other molecules.[121,122] Recently, potentiometric methods have been focused on the development of selective surfactant sensors with no sample pretreatment (even if samples are colored or viscous)[123–125] and simple measuring methods.[126] In the last few years, ISEs for the determination of ASs have been modified with nanomaterials (NMs) and carbon powders. NMs are particularly suitable for modifying ISEs because they enhance electrode response characteristics.[127] A few review papers published in the year 2000 discuss different types of sensors for the determination of various groups of surfactants.[128–130] Potentiometric sensors exclusively for ASs have been discussed only by Sanchez and del Valle.[131] In this review, special attention was given to the papers on potentiometric sensors for determination of ASs that were published after 2000.

Potentiometric surfactant sensors are used mainly as titration end-point sensors, although some direct potentiometric sensors have also been reported. The most popular sensors in this group are the coated wire-type, liquid membrane and ISFET (ion-selective field effect transistor) sensors. The sensors are suitable for determination of ionic (cationic and anionic), as well as nonionic surfactants. They can be used discontinuously in batch mode or continuously in flow injection analysis (FIA) mode.

3.4.3 Flow Injection Analysis Surfactant Sensors

Many potentiometric surfactant sensors, as well as amperometric ones, can be used as detectors in flow injection analysis (FIA). An electrochemical sensor based on the incorporation of dodecyl sulfate (DS) into polypyrrole (Ppy) by electropolymerization of pyrrole in the presence of surfactant has been described. The physicochemical characteristics of the Ppy–DS electrode and its response against alkali metal ions in cyclic voltammetry were studied. Application of this electrode as a sensor in flow-injection analysis and in high-performance ion chromatography has been demonstrated. The electrode remains stable for several weeks with no evidence of chemical or mechanical damage.[132]

The development of high-performance paraquat-selective potentiometric sensors based on octamethyl cyclotetrasiloxane has been described. In the fabrication of these sensors, the use of highly lipophilic PVC-based membrane components, namely plasticizer and anionic sites, was emphasised. The use of bis(1-butylpentyl)decane-1,1-diyl-diglutarate (BBDG) and tetra-*n*-undecyl 3,3′,4,4′-benzophenone tetracarboxylate (TBT) as plasticizers in conjunction with either sodium tetrakis3,5-bis(trifluoromethyl)phenyl borate (NTB) or potassium tetrakis(4-chlorophenyl)borate (KTPB) as membrane additives all yielded functional paraquat-selective chemical sensors with Nernstian slopes. A distinct advantage of the paraquat sensors described here was their resistance to fouling by surfactants as compared to earlier sensors that used *o*-nitrophenyl octyl ether (NPOE) as a plasticiser. The second part of this paper dealt with the use of the promising membrane systems described above when applied on the surface of platinum wire and subsequently used as a flow-through potentiometric detector. A brief description of the flow-through detector design and optimization of the flow-injection analysis (FIA) set-up was also presented. Analysis of 0.01–0.1 mM paraquat dichloride in the backgrounds of well, river, and lake waters when performed in the FIA mode gave excellent recoveries. A throughput of 85 samples per h could be achieved with this system.[133] An automated FIA system for the determination of low levels of anionic surfactants in river water and wastewater was developed. The system uses specially constructed tubular flow-through ion-selective electrodes (ISEs) as potentiometric sensors with on-line preconcentration techniques. The anionic surfactant ISEs employed were of the all-solid-state type with a plasticized PVC membrane. They showed a general response to anionic surfactants with a lower limit of linear response of approximately 10^{-5} M, when used in direct determinations. However, their specificity was limited, which hampered their direct use with environmental samples. Therefore, the FIA system presented here included a solid-phase extraction procedure for the purification and preconcentration of analytes. The FIA system was first applied for the determination of different types of anionic surfactant standards. Potentially interfering substances such as chloride, nitrate, and nonionic surfactants were checked to verify that they did not interfere with the response of the system. Concentrations of approximately 10^{-7} M (0.03 ppm) of sodium dodecyl sulfate could be detected in the non-linear response region when 3 mL of sample was preconcentrated and eluted with 50 L of acetonitrile in water with a volume fraction of 75%. Precision was 2% RSD ($N = 31$) for 10^{-6} M sodium dodecyl sulfate standard solution and the sample throughput was 10 h^{-1}. The FIA system was then used for the determination of total anionic surfactants in river water and wastewater.[134]

3.4.4 Amperometric Surfactant Sensors

Among amperometric surfactant sensors there are several true surfactant biosensors that employ microorganisms or cells. A range of modified PVC membranes based on their permeability and selectivity to electrochemically active species has been characterized. The species were detected at an amperometric (two-electrode) cell in phosphate buffer solution. Diffusing species studied were paracetamol, catechol, ascorbic acid, and hydrogen peroxide. Membrane-modifying agents studied were cationic (Aliquat 336s) and nonionic (Tween 80 and Triton X-100) surfactants. For the first time, surfactant combinations with liquid crystal *N*-(4-methoxybenzylidene)-4-butyl-aniline were reported.[135]

Bacteria-degraders as the base of an amperometric biosensor for detection of anionic surfactants have been investigated. A *Pseudomonas rathonis* T-based amperometric biosensor was constructed for the detection of anionic surfactants. The microorganisms contained the plasmid for surfactant degradation. The sensor was highly sensitive to sodium dodecyl sulfate (SDS) and volgonat. The lower limit of SDS detection was within the range 0.25–0.75 mg L^{-1}.[136] An amperometric sensor study of the selectivity of PVC membranes plasticized with surfactants and liquid crystals was reported.[137] The effects of nonionic surfactants on the electrochemical behavior of ubiquinone and menaquinone incorporated in a carbon paste electrode were studied. A carbon paste electrode, in which the carbon particles were coated with a thin layer of a nonionic surfactant, was constructed with a pasting liquid containing ubiquinone or menaquinone. Further, such a modification was applied for the preparation of an enzyme electrode in which the quinone molecule acts as a redox mediator and influences on the sensitivity of the glucose biosensor were demonstrated.[138] The selective binding of charge-diffuse alkyl and arylammonium ions relies on multiple weak interactions with a complementary synthetic receptor. Using appropriately sized lipophilic cyclodextrin derivatives, the chemoselective binding of alkylammonium ions such as dopamine, acetyl choline, guanidine, and long-chain cationic surfactants may be achieved, allowing their selective detection by either potentiometric or amperometric methods of analysis. Enantioselectivity in the binding of chiral hydroxyarylammonium ions, such as propranolol, allows chiral sensors to be developed. The selective detection of various clinically important analytes, such as imipramine, lignocaine, and creatinine, has also been studied.[139] The applicability of the well-known three-parameter Hill equation for the description of calibration curves for potentiometric (detection of glucose, pesticides, urea) and amperometric (detection of surfactants, biphenyl, nitrite) biosensors has been analyzed. A criterion for validity of the Hill equation has been proposed. The sources of errors in the determination of concentration of the analyzed substance have been considered. Ordered surfactant films on electrodes are promising for various applications, including catalysis, sensors, and fundamental studies. Films were prepared by first casting didodecyldimethylammonium bromide (DDAB) onto electrodes, then introducing copper phthalocyanine tetrasulfonate (CuPCTS) by ion exchange from solution, or by casting the films as the CuPCTS (didodecyldimethylammonium) salt. The salt films were characterized by overlapped voltammetric peaks, while the ion-exchanged films showed a well-resolved first reduction process. The analysis of voltammetric data at $pH = 12$ showed that charge-transport diffusion and electron-transfer kinetics were significantly faster in the ion-exchanged films than in the salt films. The fluid lamellar liquid crystal structure of the ion-exchanged films facilitated charge transport compared to the salt films, which are not liquid crystalline.[140]

3.4.5 Conductometric Surfactant Sensors

There is only one paper dealing with this type of surfactant sensor for on-line monitoring of the evolution of a number of particles in an emulsion polymerization by conductivity measurements. The model developed was built on the assumption that the surfactant is partitioned among the three principal phases of the polymerizing latex. A soft-sensor strategy was then proposed for monitoring the number of polymer particles (N-P) by combining the conductivity model with the available conversion, temperature, and conductivity signals. The main objective was to validate the conductivity model in a broader range of operating conditions and to follow the evolution of N-P (nucleation/coagulation) on-line under different reaction conditions. A series of batch and semi-batch polymerization runs was carried out aiming to evaluate the model performance when different solid content, surfactant concentration, and reactor temperatures are used. Results showed that the model was able to perform accurate predictions of N-P, even when disturbances of 2 °C in temperature and 0.03 in monomer conversion took place during the polymerization process.[140]

3.4.6 Impedimetric Surfactant Sensors

Simple methods of preparing direct affinity sensors have been proposed. Due to the self-consistent introduction of a hydrocarbon chain bound using oligonucleotide pentadecathymidylate into the hydrophobic region of surfactant bilayer or the adsorption of antibodies on the bilayer surface, the immobilizations of oligonucleotides or antibodies were correspondingly carried out. The responses were detected by impedance spectroscopy. Whereas specific DNA-coupling caused a decrease in the real part of the impedance, the antibody–antigen interaction caused its increase. The obtained results give an opportunity for the development of impedimetric affinity sensors for clinical analysis or for the detection of various environmental pollutants.[141,142]

A few hundred ppm of water can cause detrimental changes in the lubricating properties of engine oil. Electrochemical sensors based on electrochemical impedance spectroscopy and cyclic voltammetry have been utilized to detect water leaks and continuously monitor the time-dependent dynamics of water–oil interactions following the injection of water into industrial lubricant. Immediately following the injection, water molecules interacted with the oil additives (surfactants) forming a water-in-oil emulsion based on inverse micelles. Emulsification was followed by gradual loss of water from the solution through evaporation and electrolysis. On-line data were used to characterize the dynamics of water micellization, evaporation, and electrolysis. The values of kinetic rate constants and diffusion coefficients for the components of the water–oil system were determined. In order to support the experimental data and establish the kinetics of water–oil interactions, literature equations describing these interactions were adopted to develop a computational analysis model. The model illustrated the processes occurring in the water–oil system and resulted in an increased understanding of the recorded experimental data.[143]

3.5 A Bio-inspired Sensor

The bio-inspired concept comes from the observation of molecules and architecture inspired by biological models. Mechanisms, functions, structures, and design features have inspired the development of new synthetic materials, such as mimics of enzyme catalytic sites, bioceramics, and biosensors, with applications in biomedical engineering and bionanotechnology. In recent years, Neves *et al.* have developed bio-inspired metallocomplexes with promiscuous character for catalytic redox reactions. An example is the complex $[(bpbpmp)Fe^{III}(m\text{-}OAc)_2\text{–}Fe^{II}](ClO_4)$, a synthetic promiscuous catalyst, which catalyzes the oxidation of catechols and the hydrolysis of phosphodiesters.[144]

Hydroquinone (HQ) and catechol (CC) are two important isomers commonly found in nature. Because of their high toxicity and low degradability they are both considered environmental pollutants by the US Environmental Protection Agency (EPA) and the European Union (EU). These compounds have several uses, for instance, as depressors, antioxidants, and dyes, and they are also found in cigarette residues. Since these isomers have similar redox properties and usually coexist in environmental samples, the development of a simultaneous, simple and fast analytical method for their detection is desirable.[145,146]

3.6 Conclusions and Outlook

Surfactants that are surface-active agents and a particular class of amphiphilic molecules are widely used in electrochemistry to improve electrode–solution interface properties. Voltammetric methods are extensively applied due to their low cost, simple preparation, speed, and outstanding analytical performance. The addition of surfactants to electrolytes or immobilizing surfactants on the electrode surface may greatly enhance the sensitivity and selectivity by changing the redox potential and charge-transfer coefficients, and increasing the electron transfer rate. Moreover, applications in surfactant-based voltammetric sensors, have been discussed. Surfactants are mostly added below their CMCs, either into the electrolyte solution or to the electrode composition. Usually, the sensitivity increases with increasing length of the hydrophobic segment of the surfactant, unless the longer electron-transfer distance for the target hinders the electron-transfer rate. Modified electrodes, along with surfactants, have been proved to show excellent electroanalytical properties, such as high surface-to-volume ratio, strong adsorptive ability, high electrical conductivity, chemical stability, wide potential range, low background current, and good biocompatibility. Surfactant-based voltammetric sensors can analyze a real sample in various areas such as biomedical, food, agriculture, and industry. Gemini surfactants promote electron transfer and show many benefits over conventional surfactants due to their unique properties, such as lower CMC, higher efficiency in lowering the surface tension of water, stronger solubilization, and surface-wetting ability. Therefore, they provide new opportunities in the construction of modified electrodes. This chapter can assist researchers to use and develop surfactant-based voltammetric sensors. In conclusion, despite the extensive research performed in this area, there are still several limitations and challenges to overcome. Future work is needed to improve the selectivity of these sensors in order to engage simply with real samples without requiring extensive sample preparation methods.

References

1. M. Jozanović, N. Sakač, M. Karnaš and M. Medvidović-Kosanović, Potentiometric Sensors for the Determination of Anionic Surfactants–A Review, *Crit. Rev. Anal. Chem.*, 2019, 115–137.
2. Global Anionic Surfactants Market—Industry Trends and Forecast to 2027; Data Bridge Market Research, Maharashtra, India, 2020.
3. D. C. Barson, Using consumer research in the development and restaging of personal care products, *Surfactants in Personal Care Products and Decorative Cosmetics*, 3rd edn, 2006, pp. 95–105.
4. ISO16265, *Water Quality—Determination of the Methylene Blue Active Substances (MBAS) Index—Method Using Continuous Flow Analysis (CFA)*, 2009. Available online: https://www.iso.org/standard/52130.html (accessed on 9 August 2018).
5. ISO, *Surface Active Agents, Detergents, Determination of Anionic-Active Matter by Manual or Mechanical Direct Two-Phase Titration Procedure*, ISO, Geneva, Switzerland, 1989.
6. S. M. Shaban, I. Aiad, M. M. El-Sukkary, E. A. Soliman and M. Y. ElAwady, One step green synthesis of hexagonal silver nanoparticles and their biological activity, *J. Ind. Eng. Chem.*, 2014, **20**, 4473–4481.
7. S. D. Wettig and R. E. Verrall, Thermodynamic Studies of Aqueous m–s–m Gemini Surfactant Systems, *J. Colloid Interface Sci.*, 2001, **235**, 310–316.
8. L.-G. Qiu, A.-J. Xie and Y.-H. Shen, A novel triazole-based cationic gemini surfactant: synthesis and effect on corrosion inhibition of carbon steel in hydrochloric acid, *Mater. Chem. Phys.*, 2005, **91**, 269–273.
9. I. Aiad, S. M. Shaban, H. Y. Moustafa and A. Hamed, Experimental Investigation of Newly Synthesized Gemini Cationic Surfactants as Corrosion Inhibitors of Mild Steel in 1.0 M HCl, *Prot. Met. Phys. Chem. Surf.*, 2018, **54**, 135–147.
10. J. Xu, J. Hu, C. Peng, H. Liu and Y. Hu, A simple approach to the synthesis of silver nanowires by hydrothermal process in the presence of gemini surfactant, *J. Colloid Interface Sci.*, 2006, **298**, 689–693.
11. Q. Liu, M. Guo, Z. Nie, J. Yuan, J. Tan and S. Yao, Spacer-Mediated Synthesis of Size-Controlled Gold Nanoparticles Using Geminis as Ligands, *Langmuir*, 2008, **24**, 1595–1599.
12. M. Jozanović, N. Sakač, M. Karnaš and M. Medvidović-Kosanović, Potentiometric sensors for the determination of anionic surfactants–a review, *Crit. Rev. Anal. Chem.*, 2019, 1–23.
13. Y. Shao, Y. Ying and J. Ping, Recent advances in solid-contact ion-selective electrodes: functional materials, transduction mechanisms, and development trends, *Chem. Soc. Rev.*, 2020, **49**, 4405–4465.
14. M. Cuartero and G. A. Crespo, All-solid-state potentiometric sensors: a new wave for in situ aquatic research, *Curr. Opin. Electrochem.*, 2018, **10**, 98–106.

15. E. Olkowska, Z. Polkowska and J. Namieśnik, Analytical procedures for the determination of surfactants in environmental samples, *Talanta*, 2012, **88**, 1–13.
16. M. Karnaš, N. Sakač, M. Jozanović, M. Tsakiri, M. Kopriva, E. K. Andrić and M. Sak-Bosnar, The influence of plasticisers on response characteristics of anionic surfactant potentiometric sensor, *Int. J. Electrochem. Sci.*, 2017, **12**, 5921–5933.
17. H. Kaur, M. Chhibber and S. K. Mittal, Acyclic arylamine-based ionophores as potentiometric sensors for Zn^{2+} and Ni^{2+} ions, *J. Carbon Res.*, 2017, **3**, 1–13.
18. M. Fizer, O. Fizer, V. Sidey, R. Mariychuk and Y. Studenyak, Experimental and theoretical study on cetylpyridinium dipicrylamide – a promising ion-exchanger for cetylpyridinium selective electrodes, *J. Mol. Struct.*, 2019, **1187**, 77–85.
19. P. L. Padnya, A. V. Porfireva, G. A. Evtugyn and I. I. Stoikov, Solid contact potentiometric sensors based on a new class of ionic liquids on thiacalixarene platform, *Front. Chem.*, 2018, **6**, 1–12.
20. M. Cuartero, J. A. Ortuño, M. S. García, G. Sánchez, M. Más-Montoya and D. Curiel, Benzodipyrrole derivates as new ionophores for anion-selective electrodes: improving potentiometric selectivity towards divalent anions, *Talanta*, 2011, **85**, 1876–1881.
21. M. Cuartero, M. Mas-Montoya, M. Soledad García, D. Curiel and J. A. Ortuno, New carbazolo[1,2-*a*]carbazole derivative as ionophore for anion-selective electrodes: remarkable recognition towards dicarboxylate anions, *Talanta*, 2014, **123**, 200–206.
22. I. A. Pechenkina and K. N. Mikhelson, Materials for the ionophore-based membranes for ion-selective electrodes: problems and achievements (review paper), *Russ. J. Electrochem.*, 2015, **51**, 93–102.
23. P. Bühlmann, E. Pretsch and E. Bakker, Carrier-based ion-selective electrodes and bulk optodes. 2. Ionophores for potentiometric and optical sensors, *Chem. Rev.*, 1998, **98**, 1593–1687.
24. P. J. Greenawalt and S. Amemiya, Voltammetric mechanism of multiion detection with thin ionophore-based polymeric membrane, *Anal. Chem.*, 2016, **88**, 5827–5834.
25. M. Sak-Bosnar, R. Matesic-Puac, D. Madunic-Cacic and Z. Grabaric, New potentiometric sensor for determination of low levels of anionic surfactants in industrial effluents, *Tenside, Surfactants, Deterg.*, 2006, **43**, 82–87.
26. D. Madunić-Čačić, M. Sak-Bosnar, R. A Matešić-Puač and Z. Grabarić, Determination of anionic surfactants in real systems using 1,3-didecyl-2-methyl-imidazolium-tetraphenylborate as sensing material, *Sens. Lett.*, 2008, **6**, 339–346.
27. L. Zelenka, M. Sak-Bosnar, N. Marek and B. Kovacs, A new anionic surfactantsensitive potentiometric sensor with a highly lipophilic electroactive material, *Int. J. Electrochem. Sci.*, 2011, **6**, 240–253.

28. G. A. E. Mostafa, PVC matrix membrane sensor for potentiometric determination of dodecylsulfate, *Int. J. Environ. Anal. Chem.*, 2008, **88**, 435–446.
29. R. K. Mahajan and A. Shaheen, Effect of various additives on the performance of a newly developed PVC based potentiometric sensor for anionic surfactants, *J. Colloid Interface Sci.*, 2008, **326**, 191–195.
30. S. Devi and M. C. Chattopadhyaya, Determination of sodium dodecyl sulfate in toothpastes by a PVC matrix membrane sensor, *J. Surfactants Deterg.*, 2013, **16**, 391–396.
31. Y. M. Issa, S. H. Mohamed and M. A. El Baset, Chemically modified carbon paste and membrane sensors for the determination of benzethonium chloride and some anionic surfactants (SLES, SDS, and LABSA): characterization using SEM and AFM, *Talanta*, 2016, **155**, 158–167.
32. I. Kowalczyk, Synthesis, molecular structure and spectral properties of quaternary ammonium derivatives of 1,1-dimethyl-1,3-propylenediamine, *Molecules*, 2008, **13**, 379–390.
33. D. Demberelnyamba, K. S. Kim, S. Choi, S. Y. Park, H. Lee, C.-J. Kim and I.-D. Yoo, Synthesis and antimicrobial properties of imidazolium and pyrrolidinonium salts, *Bioorg. Med. Chem.*, 2004, **12**, 853–857.
34. B. Brycki, I. Kowalczyk and A. Kozirog, Synthesis, molecular structure, spectral properties and antifungal activity of polymethylene-α,ω-bis(*n*,*n*-dimethyl-ndodecyloammonium bromides), *Molecules*, 2011, **16**, 319–335.
35. B. Niess and K. A. Jørgensen, The asymmetric vinylogous Mannich reaction of dicyanoalkylidenes with α-amido sulfones under phase-transfer conditions, *Chem. Commun.*, 2007, 1620–1622.
36. T. Fujimori, K. Fujii, R. Kanzaki, K. Chiba, H. Yamamoto, Y. Umebayashi and S. Ishiguro, Conformational structure of room temperature ionic liquid N-butyl-Nmethyl-pyrrolidinium bis(trifluoromethanesulfonyl) imide – Raman spectroscopic study and DFT calculations, *J. Mol. Liq.*, 2007, **131–132**, 216–224.
37. D. M. Tigelaar, M. A. B. Meador and W. R. Bennett, Composite electrolytes for lithium batteries: Ionic liquids in APTES cross-linked polymers, *Macromolecules*, 2007, **40**, 4159–4164.
38. B. Paczosa-Bator, R. Piech and A. Lewenstam, Determination of the leaching of polymeric ion-selective membrane components by stripping voltammetry, *Talanta*, 2010, **81**, 1003–1009.
39. M. Hajduković, M. Samardžić, O. Galović, A. Széchenyi and M. Sak-Bosnar, A functionalized nanomaterial based, new, solid state cationic-surfactant-selective sensor with fast response and low noise, *Sens. Actuators, B*, 2017, **251**, 795–803.
40. K. J. Miller, Additivity methods in molecular polarizability, *J. Am. Chem. Soc.*, 1990, **112**, 8533–8542.
41. Molecular Interactions (Noncovalent Interactions), (n.d.). https://ww2.chemistry.gatech.edu/~lw26/structure/molecular_interactions/mol_int.html (accessed 7 December 2020).

42. M. Samardžić, M. Budetić, A. Széchenyi, D. Marković, P. Živković, B. Šarkanj and M. Jozanović, The novel anionic surfactant selective sensors based on newly synthesized quaternary ammonium salts as ionophores, *Sens. Actuators, B*, 2021, **343**, 130103.
43. N. Sakač, D. Marković, B. Šarkanj, D. Madunić-Čačić, K. Hajdek, B. Smoljan and M. Jozanović, Direct Potentiometric Study of Cationic and Nonionic Surfactants in Disinfectants and Personal Care Products by New Surfactant Sensor Based on 1,3-Dihexadecyl-1*H*-benzo[*d*]imidazol-3-ium, *Molecules*, 2021, **26**, 1366.
44. C. Gavach and C. Bertrand, Electrodes Specifiques D'anions a Longue Chaine Hydro-Carbonee: Application Au Dosage Potentiometrique de Detergents Anioniques, *Anal. Chim. Acta*, 1971, **55**, 385–393.
45. B. J. Birch and D. E. Clarke, An Electrode Selective to the Dodecyl Sulphate Anion, *Anal. Chim. Acta*, 1972, **61**, 159–162.
46. N. Ciocan and D. F. Anghel, An Ion-Extractive Liquid-Membrane Anionic Surfactant Sensitive Electrode and Its Analytical Applications, *Fresenius' Z. Anal. Chem.*, 1978, **290**, 237–240.
47. W. Szczepaniak and M. Ren, Selectivity of Liquid Membrane Electrode Based on Mercurated Polystyrene as Ion-Exchanger to Anionic Surfactants and Soaps, *Electroanalysis*, 1994, **6**, 341–347.
48. G. C. Kresheck and I. Constantinidis, Ion-Selective Electrodes for Octyl and Decyl Sulfate Surfactants, *Anal. Chem.*, 1984, **56**, 152–156.
49. L. Campanella, F. Mazzei and M. Tomassetti, Determinazione Di Tensioattivi Anionici Nelle Acque Mediante l'uso Di Un Elettrodo Iono-Selettivo, *Inquinamento*, 1987, **6**, 44–46.
50. M. Arvand-Barmchi, M. F. Mousavi, M. A. Zanjanchi and M. A. Shamsipur, New Dodecylsulfate-Selective Supported Liquid Membrane Electrode Based on Its N-Cetylpyridinium Ion-Pair, *Microchem. J.*, 2003, **74**, 149–156.
51. E. Pungor, The New Theory of Ion-Selective Electrodes, *Anal. Sci.*, 1998, **14**, 249–256.
52. E. Khalaed, N. A. Hassan, M. A. Abdelaziz and R. O. El-Attar, Novel Enzymatic Potentiometric Approaches for Surfactant Analysis, *Electroanalysis*, 2016, **29**, 716–721.
53. A. Craggs, G. Moody and J. Thomas, PVC Matrix Membrane Ion-Selective Electrodes, *J. Chem. Educ.*, 1974, **51**, 541–544.
54. C. Mihali and N. Vaum, Use of Plasticizers for Electrochemical Sensors, in *Recent Advances in Plasticizers*, ed. M. Luqman, InTech, Rijeka, 2012, pp. 125–140.
55. Y. Temerk, H. Ibrahim and N. Farhan, Square wave adsorptive stripping voltammetric determination of anticancer drug nilutamide in biological fluids using cationic surfactant cetyltrimethylammonium bromide, *Anal. Methods*, 2015, **7**, 9137–9144.
56. S. M. Azab, M. Shehata and A. M. Fekry, A novel electrochemical analysis of the legal psychoactive drug caffeine using a zeolite/MWCNT modified carbon paste sensor, *New J. Chem.*, 2019, **43**, 15359–15367.

57. C. Barbero, M. C. Miras, B. Schnyder, O. Hass and R. Kötz, Sulfonated polyaniline films as cation insertion electrodes for battery applications. Part 1.—Structural and electrochemical characterization, *J. Mater. Chem.*, 1994, **4**, 1775–1783.
58. S. Cinti, D. Neagu, M. Carbone, I. Cacciotti, D. Moscone and F. Arduini, Novel carbon black-cobalt phthalocyanine nanocomposite as sensing platform to detect organophosphorus pollutants at screen-printed electrode, *Electrochim. Acta*, 2016, **188**, 574–581.
59. R. A. Dar, P. K. Brahman, S. Tiwari and K. S. Pitre, Electrochemical studies of quinine in surfactant media using hanging mercury drop electrode: a cyclic voltammetric study, *Colloids Surf., B*, 2012, **98**, 72–79.
60. V. M. De Menezes, A. R. Rocha, I. Zanella, R. Mota, A. Fazzio and S. B. Fagan, Electronic transport properties of ascorbic acid and nicotinamide adsorbed on single-walled carbon nanotubes, *Chem. Phys. Lett.*, 2011, **506**, 233–238.
61. D. K. Fogliatto, A. M. J. Barbosa and V. S. Ferreira, Voltammetric determination of the phytoestrogen genistein in soy flours and soy based supplements using cationic surfactant cetyltrimetylammonium bromide, *Colloids Surf., B*, 2010, **78**, 243–249.
62. Y. Gao, T. Sun, C. Yang and B. Zhou, A sensitive electrochemical method for detection of histone deacetylase with the signal amplification of ferrocene-capped gold nanoparticle/streptavidin conjugates, *Int. J. Electrochem. Sci.*, 2015, **10**, 3230–3235.
63. H. S. El-Desoky, M. M. Ghoneim and M. M. Abdel-Galeil, Stripping voltammetric methods for determination of the antiparasitic drug nitazoxanide in bulk form, pharmaceutical formulation and human serum, *J. Braz. Chem. Soc.*, 2010, **21**, 669–679.
64. G. A. Evtugyn, R. V. Shamagsumova, P. V. Padnya, I. I. Stoikov and I. S. Antipin, Cholinesterase sensor based on glassy carbon electrode modified with Ag nanoparticles decorated with macrocyclic ligands, *Talanta*, 2014, **127**, 9–17.
65. V. K. Gupta, R. Jain, K. Radhapyari, N. Jadon and S. Agarwal, Voltammetric techniques for the assay of pharmaceuticals – a review, *Anal. Biochem.*, 2011, **408**, 179–196.
66. N. Hareesha, J. G. G. Manjunatha, C. Raril and G. Tigari, Design of novel surfactant modified carbon nanotube paste electrochemical for the sensitive investigation of tyrosine as a pharmaceutical drug, *Adv. Pharm. Bull.*, 2019, **9**, 132–137.
67. E. Hatami, N. Ashraf and M. H. Arbab-Zavar, On-chip electrochemical sensing of propylparaben as an endocrine disruptor model using a disposable gold platform modified with phosphomolybdate doped polypyrrole and nanodiamond, *J. Electrochem. Soc.*, 2019, **166**, B1379–B1386.
68. G. Inzelt, M. Pineri, J. W. Schultze and M. A. Vorotyntsev, Electron and proton conducting polymers: recent developments and prospects, *Electrochim. Acta*, 2000, **45**, 2403–2421.

69. C. B. Jacobs, M. J. Peairs and B. J. Venton, Carbon nanotube based electrochemical sensors for biomolecules, *Anal. Chim. Acta*, 2010, **662**, 105–127.
70. R. Jain, V. K. Gupta, N. Jadon and K. Radhapyari, Voltammetric determination of cefixime in pharmaceuticals and biological fluids, *Anal. Biochem.*, 2010, **407**, 79–88.
71. A. Levent, A. Altun, Y. Yardım and Z. Şentürk, Sensitive voltammetric determination of testosterone in pharmaceuticals and human urine using a glassy carbon electrode in the presence of cationic surfactant, *Electrochim. Acta*, 2014, **128**, 54–60.
72. A. Levent, Y. Yardım and Z. Şentürk, Electrochemical performance of boron-doped diamond electrode in surfactant-containing media for ambroxol determination, *Sens. Actuators, B*, 2014, **203**, 517–526.
73. L. Li, E. Liu, X. Wang, J. Chen and X. Zhang, Simultaneous determination of naphthol isomers at poly (3-methylthiophene)-nano-Au modified electrode with the enhancement of surfactant, *Mater. Sci. Eng., C*, 2015, **53**, 36–42.
74. W. Li, T. Luo, Y. Yang, X. Tan and L. Liu, Formation of controllable hydrophilic/hydrophobic drug delivery systems by electrospinning of vesicles, *Langmuir*, 2015, **31**, 5141–5146.
75. Q. Lu, W. Zhang, Z. Wang, G. Yu, Y. Yuan and Y. Zhou, A facile electrochemical sensor for nonylphenol determination based on the enhancement effect of cetyltrimethylammonium bromide, *Sensors*, 2013, **13**, 758–768.
76. J. G. Manjunatha, A surfactant enhanced graphene paste electrode as an effective electrochemical sensor for the sensitive and simultaneous determination of catechol and resorcinol, *Chem. Data Collect.*, 2020, **25**, 100331.
77. A. Munir, A. Shah, J. Nisar, M. N. Ashiq, M. S. Akhter and A. H. Shah, Selective and simultaneous detection of Zn^{2+}, Cd^{2+}, Pb^{2+}, Cu^{2+}, Hg^{2+} and Sr^{2+} using surfactant modified electrochemical sensors, *Electrochim. Acta*, 2019, **323**, 134592.
78. F. Nasiri, G. H. Rounaghi, N. Ashraf and B. Deiminiat, A new electrochemical sensing platform for quantitative determination of diclofenac based on gold nanoparticles decorated multiwalled carbon nanotubes/graphene oxide nanocomposite film, *Int. J. Environ. Anal. Chem.*, 2021, **101**, 153–166.
79. P. Nunes Angelis, J. de Cassia Mendonça, L. Rianne da Rocha, T. Boareto Capelari, M. Carolyne Prete and M. Gava Segatelli, *et al.*, Feasibility of a nano-carbon black paste electrode for simultaneous voltammetric determination of antioxidants in food samples and biodiesel in the presence of surfactant, *Electroanalysis*, 2020, **32**, 1198–1207.
80. G. Ozcelikay, B. Dogan-Topal and S. A. Ozkan, An electrochemical sensor based on silver nanoparticles-benzalkonium chloride for the voltammetric determination of antiviral drug tenofovir, *Electroanalysis*, 2018, **30**, 943–954.

81. S. G. Pawar, S. L. Patil, M. A. Chougule, A. T. Mane, D. M. Jundale and V. B. Patil, Synthesis and characterization of polyaniline: TiO_2 nanocomposites, *Int. J. Polym. Mater.*, 2010, **59**, 777–785.
82. S. S. Shankar, B. K. Swamy, B. N. Chandrashekar and K. J. Gururaj, Sodium do-decyl benzene sulfate modified carbon paste electrode as an electrochemical sensor for the simultaneous analysis of dopamine, ascorbic acid and uric acid: a voltammetric study, *J. Mol. Liq.*, 2013, **177**, 32–39.
83. D. Talarico, F. Arduini, A. Amine, D. Moscone and G. Palleschi, Screenprinted electrode modified with carbon black nanoparticles for phosphate detection by measuring the electroactive phosphomolybdate complex, *Talanta*, 2015, **141**, 267–272.
84. C. F. Valezi, E. H. Duarte, G. R. Mansano, L. H. Dall'Antonia, C. R. T. Tarley and E. R. Sartori, An improved method for simultaneous square-wave voltammetric determination of amlodipine and enalapril at multiwalled carbon nanotubes paste electrode based on effect of cationic surfactant, *Sens. Actuators, B*, 2014, **205**, 234–243.
85. G.-J. Yang, X.-L. Qu, M. Shen, C.-Y. Wang, Q.-S. Qu and X.-Y. Hu, *Sens. Actuators, B*, 2007, **128**, 258–265.
86. Y. Xu, C. Hu and S. Hu, *Bioelectrochemistry*, 2009, **74**, 254–259.
87. L. Gonzales-Macia, M. R. Smytha, A. Morrin and A. J. Killard, *Electrochim. Acta*, 2011, **56**, 4146–4153.
88. J. F. Rusling and A.-E. F. Nassar, *J. Am. Chem. Soc.*, 1993, **115**, 11891–11897.
89. S. Y. Rhieu, D. R. Ludwig, V. S. Siu, G. Tayhas and R. Palmore, *Electrochem. Commun.*, 2009, **11**, 1857–1860.
90. J. Li, L. Zhou, X. Han and H. Liu, *Sens. Actuators, B*, 2008, **135**, 322–326.
91. W. Shan, H. Liu, J. Shi, L. Yang and N. Hu, *Biophys. Chem.*, 2008, **134**, 101–109.
92. C.-N. Chen, C.-T. Huang, W. J. Tseng and M.-H. Wei, *Appl. Surf. Sci.*, 2010, **257**, 650–655.
93. Y. Hu, N. Hu and Z. Zeng, *Talanta*, 2000, **50**, 1183–1195.
94. B. S. Souza, E. C. Leopoldino, D. W. Tondo, J. Dupont and F. Nome, *Langmuir*, 2012, **28**, 833–840.
95. J. Eastoe, M. J. Hollamby and L. Hudson, *Adv. Colloid Interface Sci.*, 2006, **128**, 5–15.
96. J. Wu, Q. Fu, B. Lu, H. Li and Z. Xu, *Thin Solid Films*, 2010, **518**, 3240–3245.
97. X. Cao, L. Luo, Y. Ding, X. Zou and R. Bian, *Sens. Actuators, B*, 2008, **129**, 941–946.
98. I. Mahmud, A. J. F. Samed, M. A. Haque and M. A. B. H. Susan, *J. Saudi Chem. Soc.*, 2011, **15**, 203–208.
99. A. Sarkar, R. Chadha, N. Biswas, T. Mukherjee and S. Kapoor, *J. Colloid Interface Sci.*, 2009, **332**, 224–230.
100. D. Mimica, J. H. Zagal and F. Bedioui, *J. Electroanal. Chem.*, 2001, **497**, 106–113.

101. G. Verma and P. A. Hassan, *Phys. Chem. Chem. Phys.*, 2013, **15**, 17016–17028.
102. M. Ramanathan, L. K. Shrestha, T. Mori, Q. Ji, J. P. Hill and K. Ariga, *Phys. Chem. Chem. Phys.*, 2013, **15**, 10580–10611.
103. P. Pallavicini, Y. A. Diaz-Fernandez and L. Pasotti, *Coord. Chem. Rev.*, 2009, **253**, 2226–2240.
104. P. Bandyopadhyay and A. K. Ghosh, *Sens. Lett.*, 2011, **9**, 1249–1264.
105. M. Jamkratoke, G. Tumcharern, T. Tuntulani and B. Tomapatanaget, *J. Fluoresc.*, 2011, **21**, 1179–1187.
106. K. Ngamdee, T. Noipa, S. Martwiset, T. Tuntulani and W. Ngeontae, *Sens. Actuators, B*, 2011, **160**, 129–138.
107. A. Mallick, M. C. Mandal, B. Haldar, A. Chakrabarty, P. Das and N. Chattopadhyay, *J. Am. Chem. Soc.*, 2006, **128**, 3126–3127.
108. S. S. Bag, R. Kundu and S. Talukdar, *Tetrahedron Lett.*, 2012, **53**, 5875–5879.
109. J. Wang, X. Qian, J. Qian and Y. Xu, *Chem. – Eur. J.*, 2007, **13**, 7543–7552.
110. H. Tian, J. Qian, H. Bai, Q. Sun, L. Zhang and W. Zhang, *Anal. Chim. Acta*, 2013, **768**, 136–142.
111. R. Hu, J. Feng, D. Hu, S. Wang, S. Li, Y. Li and G. Yang, *Angew. Chem., Int. Ed.*, 2010, **49**, 4915–4918.
112. M. Jamkratoke, V. Ruangpornvisuti, G. Tumcharern, T. Tuntulani and B. Tomapatanaget, *J. Org. Chem.*, 2009, **74**, 3919–3922.
113. A. D. Hughes, I. C. Glenn, A. D. Patrick, A. Ellington and E. V. Anslyn, *Chem. – Eur. J.*, 2008, **14**, 1822–1827.
114. E. N. Savariar, S. Ghosh, D. C. Gonzalez and S. Thayumanavan, *J. Am. Chem. Soc.*, 2008, **130**, 5416–5417.
115. M. A. Palacios, Z. Wang, V. A. Montes, G. V. Zyryanov and P. Anzenbacher, *J. Am. Chem. Soc.*, 2008, **130**, 10307–10314.
116. M. Shellaiah, Y.-H. Wu, A. Singh, M. V. R. Raju and H.-C. Lin, *J. Mater. Chem. A*, 2013, **1**, 1310–1318.
117. H. N. Kim, W. X. Ren, J. S. Kim and J. Yoon, *Chem. Soc. Rev.*, 2012, **41**, 3210–3244.
118. A. P. D. Silva, H. Q. N. Gunaratne, T. Gunnlaugsson, A. J. M. Huxley, C. P. McCoy, J. T. Rademacher and T. E. Rice, *Chem. Rev.*, 1997, **97**, 1515–1566.
119. R. Martínez-Máñez and F. Sancenón, *Chem. Rev.*, 2003, **103**, 4419–4476.
120. J. P. Anzenbacher, P. Lubal, P. Bucek, M. A. Palacios and M. E. Kozelkova, *Chem. Soc. Rev.*, 2010, **39**, 3954–3979.
121. S. Rochat, J. Gao, X. Qian, F. Zaubitzer and K. Severin, *Chem. – Eur. J.*, 2010, **16**, 104–113.
122. (a) G. V. Zyryanov, M. A. Palacios and P. Anzenbacher Jr, *Angew. Chem., Int. Ed.*, 2007, **46**, 7849–7852; (b) Z. Wang, M. A. Palacios and P. Anzenbacher Jr, *Anal. Chem.*, 2008, **80**, 7451–7459.
123. A. Volkov, E. Gorbova, A. Vylkov, D. Medvedev, A. Demin and P. Tsiakaras, Design and Applications of Potentiometric Sensors Based

on Proton-Conducting Ceramic Materials. A Brief Review, *Sens. Actuators, B*, 2017, **244**, 1004–1015.

124. E. Bakker and E. Pretsch, Nanoscale Potentiometry, *TrAC, Trends Anal. Chem.*, 2008, **27**, 612–618.
125. N. M. Makarova and E. G. Kulapina, New Potentiometric ScreenPrinted Sensors for Determination of Homologous Sodium Alkylsulfates, *Sens. Actuators, B*, 2015, **210**, 817–824.
126. G. A. Alvarez-Romero, A. Morales-Perez, A. Rojas-Hernandez, M. Palomar-Pardave and M. T. Ramırez-Silva, Development of a Tubular Sensor Based on a Polypyrrole-Doped Membrane for the Potentiometric Determination of the Dodecylsulfate Anion in a FIA System, *Electroanalysis*, 2004, **16**, 1236–1243.
127. S. Martınez-Barrachina and M. del Valle, Use of a Solid-Phase Extraction Disk Module in a FI System for the Automated Preconcentration and Determination of Surfactants Using Potentiometric Detection, *Microchem. J.*, 2006, **83**, 48–54.
128. S. S. M. Hassan, I. H. A. Badr and H. S. M. Abd-Rabboh, Potentiometric Flow Injection Analysis of Anionic Surfactants in Industrial Products and Wastes, *Microchim. Acta*, 2004, **144**, 263–269.
129. T. Masadome, S. Kugoh, M. Ishikawa, E. Kawano and S. I. Wakida, Polymer Chip Incorporated with Anionic SurfactantISFET for Microflow Analysis of Anionic Surfactants, *Sensors Actuators B Chem.*, 2005, **108**, 888–892.
130. E. M. El-Nemma, N. M. Badawi and S. S. M. Hassan, Cobalt Phthalocyanine as a Novel Molecular Recognition Reagent for Batch and Flow Injection Potentiometric and Spectrophotometric Determination of Anionic Surfactants, *Talanta*, 2009, **78**, 723–729.
131. G. Li, H. Ma and J. Hao, Surfactant Ion-Selective Electrodes: A Promising Approach to the Study of the Aggregation of Ionic Surfactants in Solution, *Soft Matter*, 2012, **8**, 896–909.
132. *Environmental Analysis by Electrochemical Sensors and Biosensors; Nanostructure Science and Technology*, ed. L. M. Moretto and K. Kalcher, Springer New York, New York, NY, 2014.
133. M. Ghaedi, S. Naderi, M. Montazerozohori, F. Taghizadeh and A. Asghari, Chemically Modified Multiwalled Carbon Nanotube Carbon Paste Electrode for Copper Determination, *Arabian J. Chem.*, 2017, **10**, S2934–S2943.
134. Y. R. Bazel, I. P. Antal, V. M. Lavra and Z. A. Kormosh, Methods for the Determination of Anionic Surfactants, *J. Anal. Chem.*, 2014, **69**, 211–236.
135. E. G. Kulapina, R. K. Chernova, N. M. Makarova and E. S. Pogorelova, Methods for Determining Synthetic Surfactants, *Rev. J. Chem.*, 2013, **3**, 323–362.
136. N. M. Makarova, Potentiometric Sensor Arrays in the Separate Determination of Synthetic Surfactant Homologs, *Rev. J. Chem.*, 2014, **4**, 21–32.

137. J. Sanchez and M. del Valle, Determination of Anionic Surfactants Employing Potentiometric Sensors – A Review, *Crit. Rev. Anal. Chem.*, 2005, **35**, 15–29.
138. R. C. Martinez, F. B. Dominguez, F. M. Gonzalez, J. H. Mendez and R. C. Orellana, *Anal. Chim. Acta*, 1993, **279**, 299–307.
139. B. Saad, M. M. Ariffin and M. I. Saleh, *Anal. Chim. Acta*, 1997, **338**, 89–96.
140. S. Martinez-Barrachina, J. Alonso, L. Matia, R. Prats and M. del Valle, *Anal. Chem.*, 1999, **71**, 3684–3691.
141. S. M. Reddy, B. Gangadharan and J. Manini, *Anal. Chim. Acta*, 2004, **508**, 71–78.
142. L. Taranova, I. Semenchuk, T. Manolov, P. Iliasov and A. Reshetilov, *Biosens. Bioelectron.*, 2002, **17**, 635–640.
143. A. N. Reshetilov, I. N. Semenchuk, P. V. Iliasov and L. A. Taranova, *Anal. Chim. Acta*, 1997, **347**, 19.
144. A. Neves, A. J. Bortoluzzi, R. Jovito, R. A. Peralta, B. Souza, B. Szpoganicz, A. C. Joussef, H. Terenzi, P. C. Severino, F. L. Fischer, G. Schenk, M. J. Riley, S. J. Smith and L. R. Gahan, *J. Braz. Chem. Soc.*, 2010, **21**, 1201–1212.
145. M. Lanznaster, A. Neves, A. J. Bortoluzzi, V. V. E. Aires, B. Szpoganicz, H. Terenzi, P. C. Severino, J. M. Fuller, S. C. Drew, L. R. Gahan, G. R. Hanson, M. J. Riley and G. A. Schenk, *J. Biol. Inorg. Chem.*, 2005, **10**, 319–332.
146. M. C. B. Oliveira, D. Mazera, M. Scarpellini, P. C. Severino, A. Neves and H. Terenzi, *Inorg. Chem.*, 2009, **48**, 2711–2713.

CHAPTER 4

An Overview of Stability, Lifetime, and Reuse of Surfactant Sensors

A. SUNILKUMAR,*[a] B. CHETHAN,[b] V. PRASAD[b] AND S. MATTEPPANAVAR[c]

[a] Department of Physics, Ballari Institute of Technology and Management, Ballari-583104, Karnataka, India; [b] Department of Physics, Indian Institute of Science, Bangalore, Karnataka, India; [c] KLE Society's BasavaprabhuKore Arts, Science and Commerce College, Chikodi-591201, Karnataka, India
*Email: sunilanegundi@gmail.com

4.1 Introduction

The word surfactant is a contraction of surface active agents. These are chemical compounds that decrease the surface tension between liquid–gas, solid–liquid, and liquid–liquid. Surfactants behave as emulsifiers, detergents, foaming agents, and wetting agents. Every molecule comprises both hydrophilic and hydrophobic properties. Hydrophilic means the molecules have a quest for water molecules and hydrophobic relates to the exclusion of water molecules.[1] Surfactants in water form aggregates based on the chemical structure; these surfactants unite the hydrophilic head and hydrophobic tail. In general, for surfactants like foaming agents and emulsifiers, the dynamics at the interfaces play a significant role in their behavior. To achieve stabilization at an interface, the adsorption and desorption of surfactants, along with surface rheology, play an important role. Surfactants can be both toxic and non-toxic in nature.[2]

Detection Science Series No. 23
Surfactant-based Sensors in Chemical and Biochemical Detection
Edited by J. G. Manjunatha

Published by the Royal Society of Chemistry, www.rsc.org

Quats, also known as quaternary ammonium cations, are made by alkylating tertiary amines with halocarbons that have long alkyl chains. The charge of the quaternary ammonium compounds is independent of the pH of their solution. There are numerous quaternary ammonium compounds, including domiphen bromide, cetrimonium, cetrimide, didecyldimethyl ammonium chloride, benzethonium chloride, methyl benzothonium chloride, cetalkonium chloride, and dofaniumchloride. These quats have profound applications as antimicrobials in detergents and disinfectants. Based on the concentration level they can cause health problems to mankind, including mild skin and respiratory irritation.[3]

Surfactants uses in daily life are depicted in Figure 4.1. Human beings depend heavily on surfactants, which are used in a variety of processes such as cleaning, dispersion, wetting, foaming, and anti-foaming. Many surfactants are used in everyday applications, such as detergents, fabric softeners, motor oils, emulsions, soaps, adhesives, and inks. Surfactants are also used in formulations of agrochemicals like insecticides, spermicides and biocides. In personal care items like toothpaste, shampoo, shower gel, hair conditioners, and cosmetics, surfactants are also used. Oils in oil wells are mobilized using alkali surfactant polymers.[4] Bio-surfactants can be used in medicinal lotions and sprays on the surface of the skin and mucous membranes. Surfactants are used with quantum dots to control their growth. Surfactants have even brought a new era in the field of nanoscience.[5]

Chemical sensors are made of various sensory parts that have an additive effect on analyte detection. The classification of different chemical sensors

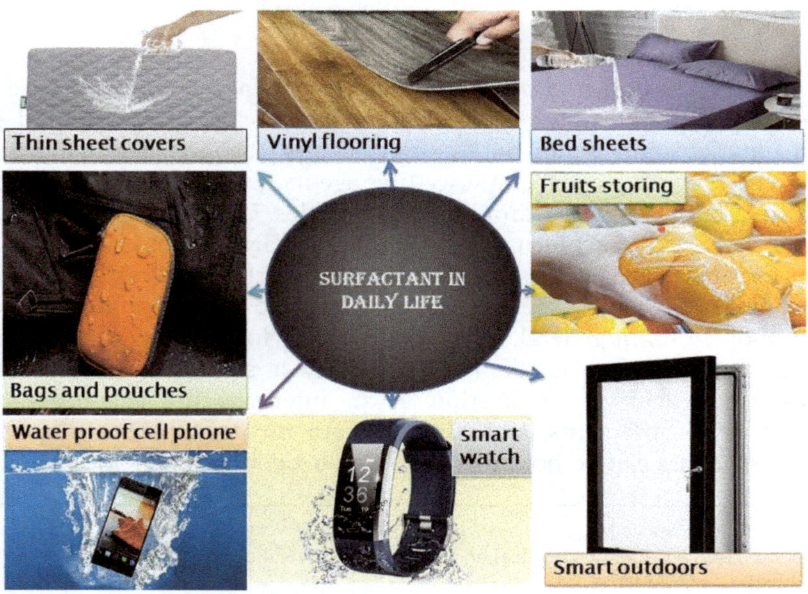

Figure 4.1 Surfactant usage in day-to-day life.

includes optical, electrical, acoustic wave, and potentiometric systems. Sensor components are chosen based on their preferred sensing characteristics, such as sensitivity, selectivity, method of detection, response and recovery time, specificity for a given class of analyte, and molecular interaction.[6] A multi-analyte test demonstrates the efficiency of the sensor, which employs various, unique sensor materials for targeted analyte detection in a mixture. Development of single-sensor technology for a target analyte based on physical, optical, and electrical characteristics comprises particular molecular targets to sustain strong and specific binding with the analyte. The performance of chemical sensors is frequently enhanced by inhibited sensing settings and changes in the ambient environment. In order to enable sensing in various combinations under all circumstances, chemical sensors work *via* an arrangement of conjunctional responses of cross-reactive sensors. A specific analyte's fingerprints are produced by combinatorial reactions to its various components. A complex mixture of phospholipids (PL) and proteins (SP) called pulmonary surfactant lowers surface tension at the air–liquid interface of the alveolus, preventing its collapse during end-expiration. Additionally, it helps the host's natural defenses against infections that are inhaled.[7]

Pneumocytes in human gestation secrete surfactant. It is mostly composed of cholesterol, phospholipids, protein, and neutral lipids. Dipalmitoylphosphatidylcholine (DPPC) phospholipid is the main surface-active agent of this surfactant. Surfactant proteins include SP-A, SP-B, SP-C, and SP-D. Surface tension is reduced and surface pressure is increased by surfactant. High surface pressure prevents the alveolar surface area from shrinking, whereas low surface tension maintains lung stability by lowering the pressure gradient of the surfactant.[8]

Olivera Galovic *et al.* have described a new graphene-based surfactant sensor for the detection of anionic surfactants in real samples. The dimethyldioctadecylammonium–tetraphenylborate (DDA–TPB) ion pair is used as a sensing material, and graphene is used as the supporting material, to create a new high-sensitivity potentiometric solid-state sensor for the measurement of anionic surfactants.[1] Dubravka Madunic *et al.* have shown the detection of cationic surfactants employing a novel sensitive potentiometric sensor in pharmaceutical disinfectants. A novel, sensitive potentiometric surfactant sensor was developed using a highly lipophilic 1,3-didecyl-2-methylimidazolium cation and a tetraphenylborate antagonist ion. This sensor was inserted into a PVC-membrane that had been plasticized as a sensing component. These sensors exhibited effective sensing applications in biochemical detection.[2] Nikola Sakac *et al.* reported a study of anionic and non-ionic surfactants in disinfectants and personal care products by direct potentiometry using a new surfactant sensor based on 1,3-dihexadecyl-1*H*-benzo[*d*]imidazol-3-ium. Results from direct potentiometric titrations of readily available cleaning products and personal care products that included cationic surfactants were in good agreement with those obtained by employing a surfactant sensor.[10,11] Hui Li *et al.* utilized acetylene black electrodes enhanced with zwitterionic surfactant and carbon nanotubes to detect triclosan in household products. A brand-new electrochemical sensor based on acetylene black, carbon nanotubes, and cetylsulfonylbetaine was developed to measure triclosan and used as a sensitizing material.[12]

There are many different mechanisms that stabilize surfactants. Use of pH buffers, chelating agents, and anti-foamers are all part of chemical stabilization. Selecting surfactant characteristics like charge, hydrophobicity, and hydrophilicity carefully is the key to achieving physical stabilization. Using surfactants and co-solvents allows microemulsions to be stabilized. The utilization of naturally occurring or biodegradable surfactants allows for the stabilization of biological processes.

Surfactants are molecules with a hydrophilic head that attracts water and a hydrophobic tail that repels it. They are employed in a number of products, including emulsifiers, detergents, and shampoos. The kind of surfactant, the concentration of surfactant, the pH of the solution, the temperature, and the presence of other substances in the solution are all variables that affect how stable a surfactant solution will be. The stability of a surfactant solution is significantly influenced by the type of surfactant used. It is crucial to select the appropriate type of surfactant for the application since some surfactants are more susceptible to degradation than others. Its stability is also influenced by the surfactant's concentration. Generally speaking, more stability will result from higher surfactant concentrations.[13]

A surfactant sensor's lifespan is influenced by the type of sensor used and the application. A surfactant sensor typically has a lifespan of several months to several years. This can be impacted by elements such as surrounding conditions, temperature, and mechanical or chemical stress.

There are many uses for recycled surfactant sensors. They can be used to control the addition of surfactants to industrial process streams, measure the concentration of surfactants in water and wastewater, monitor the effectiveness of surfactant-based treatments of contaminated soils, detection, and tracking the presence of oil and grease in air and water. They can be used in agriculture to identify pesticide presence, track the concentration of surfactants in agricultural runoff, and assess the efficacy of surfactant-based crop disease treatments. They can be used in the medical industry to gauge the amount of pharmaceutical surfactants present and to track how well respiratory disorders are being treated with surfactant-based therapies. They might be utilized in industrial settings to keep an eye out for dangerous surfactants in the streams of industrial process.[14]

This chapter on surfactant sensors focuses more on the design and development of highly stable, enhanced lifespan and reusable sensors for chemical and biochemical detection. Some examples of fascinating research fields that support the field of surfactant sensors include optical, thermal, and surface acoustic wave sensors. Chemical sensor research appears to have undergone a significant shift recently. It is among the most productive, interesting, and interdisciplinary analytical chemistry research topics. In addition, it seems to be one of the most experimental disciplines right now, where fresh ideas and thoughts have been recently explored.[12]

4.2 Classification of Surfactants

Surfactants are almost identical in nature, and comprise a hydrocarbon chain which can be framed as linear, branched, or aromatic in nature. Fluorocarbon and siloxane surfactants have identical chains and many surfactants consist of polyether chains usually ending with polar anionic groups. These materials enhance the hydrophilic property of the surfactants. Polypropylene oxides contrarily enhance the lipophilic property of a surfactant. The tail comprises the chain formation of the surfactant molecule. Based on their solubility, surfactants are mainly categorized into hydrophilic and hydrophobic, which are soluble in water and lipids respectively. Most ionic surfactants are usually water soluble.[13] Surfactant types and their general applications are shown in Figure 4.2.

Surfactants are classified into four types based on their charge of their polar head group, generally:

1. Anionic surfactants
2. Cationic surfactants
3. Non-ionic surfactants
4. Amphoteric surfactants

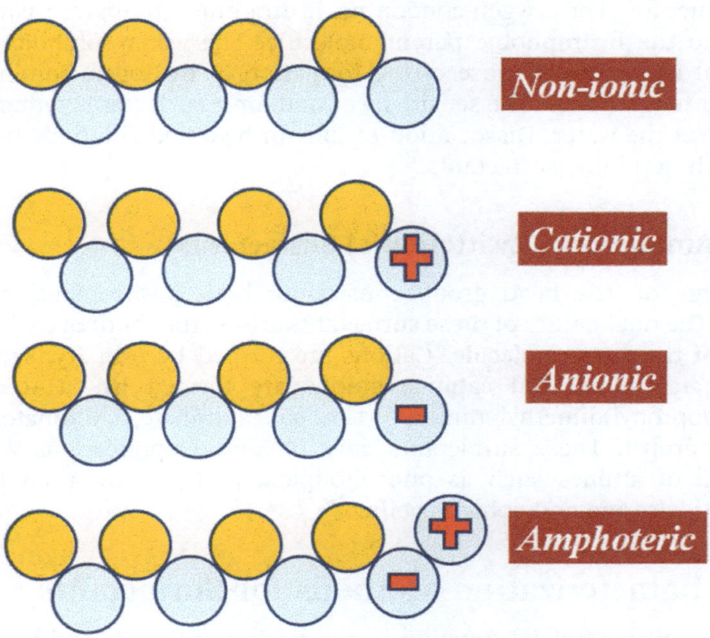

Figure 4.2 Classification of surfactants based on their composition.

4.2.1 Anionic Surfactants

The charge of the head group is generally negative and such a group comprises of sulfates, sulfonates, phosphates, and carboxylates. The most important alkyl sulfates are sodium lauryl sulfate, ammonium lauryl sulfate, sodium myreth sulfate and alkyl ether sulfates.[14] Sodium stearate incorporated in carboxylate salts produces perflorooctanaesulfonate, Doclsate along with special characteristics in sodium lauryl sarcosinate and carboxylate-based fluorosurfactants.[15] The general applications of these surfactants are shampoos and dishwashing liquids. The most common surfactants include alkyl ether sulfates, soaps, alkyl benzene sulfonates, and alkyl sulfates.

4.2.2 Cationic Surfactants

The charge of the head group is generally positive, which exhibits pH-based surfactant properties. Cationic surfactants possess adsorption behavior on charged surfaces. Most cationic surfactants comprise quaternary ammonium compounds. These surfactants can be used in softening agents, hydrophobic agents, and for hydrotropism and antistatic applications.[16]

4.2.3 Non-ionic Surfactants

These surfactants are formed by enclosing the non-charged head group. Formation of a covalent bond between the surfactants leads to a change in its structure. The oxygen-containing hydrophilic groups are covalently bonded to the hydrophobic parent molecule. The water solubility of the surfactant is the direct cause of the formation of hydrogen bonding at a particular temperature. The sensitivity of non-ionic surfactants reduces with hardness of the water. Dissociation of ions in hydrated solutions does not occur with non-ionic surfactants.[17]

4.2.4 Amphoteric (Zwitterionic) Surfactants

The charge on the head group consists of both positive and negative charges. The dual nature of these surfactants arises from both of the charged centers of the same molecule. Cations are formed by primary, secondary and ternary ammonium cations; anions are formed by CHAPS (3-[(3-cholamidopropyl)dimethylammonio]-1-propanesulfonate, sulfonates, and cocamidopropyl. These surfactants have biological applications with the formation of amines such as phospholipids, phosphatidylethanolamine, phosphatidylserine and sphigomyelins.[18]

4.3 Characterization Methods for Surfactants

Surfactants play a crucial function in the mixing of liquids and when they come into contact with solid materials as wetting agents, emulsifiers,

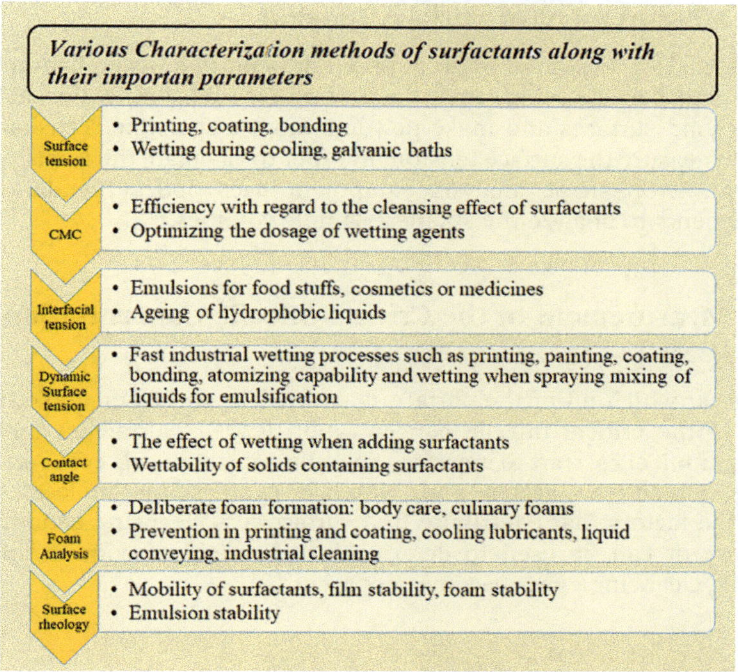

Figure 4.3 Various characterization methods and their uses.

cleansers, or foaming agents. Figure 4.3 shows the various characterization methods for surfactants along with their applications. The creation, optimization, and proper dosage of surfactants are usually thorough analysis of the interface. We can assess the effectiveness and efficiency of surfactants using tensiometers, which measure surface tension. They aid in the construction and optimization of emulsions by precisely measuring the interfacial tension. The knowledge required to use surfactants in quick processes is provided by dynamic tensiometers. The degree of wetness can be accurately determined by analyses performed with contact angle measurement equipment. Furthermore, a foamy product can be optimized or foam avoided with the use of foam analysis tools.[19] The different types of characterization methods for surfactants are:

- Measurement of surface tension.
- Measurement of the critical micelle concentration (CMC).
- Measurement of the interfacial tension between liquids.
- Measurement of dynamic surface tension and interfacial tension.
- Measurement of contact angle.
- Foam analysis.
- Surface rheological investigations.

4.3.1 Measurement of Surface Tension

The high surface tension of water makes it difficult to moisten hydrophobic surfaces and particles. Surfactants reduce surface tension, which helps to moisten solid surfaces and make powders more dispersible. Tensiometers precisely measure the surface tension, in order to calculate the impact of the surfactant. Drop shape analysis equipment uses optical pendant drop measurements to analyze the smallest sample volumes.[20]

4.3.2 Measurement of the Critical Micelle Concentration (CMC)

The level at which all of the surface is covered in surfactant molecules is known as the critical micelle concentration (CMC). Surfactant molecule clusters, or micelles, start to appear in the solution above this concentration. Micelles, which can accept and mobilize hydrophobic molecules, are among the factors that contribute to a surfactant's cleansing action. CMC measurement can be used to determine the surfactant concentration at which the cleansing effect becomes apparent.[21]

4.3.3 Measurement of the Interfacial Tension Between Liquids

Surfactants act as emulsifiers by lowering the interfacial tension between incompatible phases and enhancing mixing. They simultaneously facilitate the formation of tiny droplets and stabilize the emulsion by slowing or preventing droplet fusion. The interfacial tension between a liquid and water determines its hydrophobic nature. The interfacial tension is therefore measured in order to assure the quality of liquids where a phase mixing with water is undesirable.[22]

4.3.4 Measurement of Dynamic Surface Tension and Interfacial Tension

Various applications and manufacturing processes involve the rapid formation of surfaces and interfaces. But the diffusion and adsorption times for surfactants are very particular. Consequently, the surface tension and interfacial tension are influenced by the amount of time that has passed for the development of a new phase boundary. As a result, with dynamic processes, in a steady state, the interfacial tension or surface tension might be significantly larger. The appropriate selection and dosage of a surfactant are substantially impacted by the circumstances. Dynamic surface tension is visible above the surface, using bubble pressure tensiometers. Printing, painting, coating, and bonding are quick industrial wetting processes.

Atomization and its ability to moisten liquid emulsification are possible through mixing.[23]

4.3.5 Measurement of Contact Angle

When a surfactant is added, the wetting of a liquid is improved to a certain extent, which is indicated by the contact angle. Depending on the requirement, the contact angle is measured by drop form analysis tools with a high degree of automation. The conditions of a wetting process can be accurately simulated by measurements at high or low temperatures, under controlled humidity, or with varying dosing dynamics.[24]

4.3.6 Foam Analysis

Surfactants regularly create foams and are frequently employed alone to do so. Specific application areas necessitate varied standards for foamability, foam stability, moisture content, and bubble size. Foam analysis tools measure the foam height with relation to time using repeatable foam generation processes and methods. A conductivity module can measure the amount of moisture in the foam as well as how the height of a foam column changes over time at various heights. For the purpose of specifically optimizing the foam's characteristics, an optical foam structure analysis module estimates the bubble size distribution and its time dependency.[25]

4.3.7 Surface Rheological Investigations

In surface rheology, the rate of change in the interface region is compared to the rate of change in surface tension or interfacial tension. Surface rheology equipment evaluates a number of factors, including the interface viscosity and elasticity. Using these numbers, inferences can be made about the stability of liquid films and droplets in emulsions, as well as the mobility of surfactants. Additionally, the outcomes are typically connected with foam properties like drainage rate and foam bubble stability.[26,28,29]

4.4 Stability of Surfactant Sensors

Surfactants are "head-and-tail" molecules that interact with other molecules in a solution. The hydrophobic/hydrophilic balance, head and tail size, charge, and other parameters affect the stability of a surfactant solution. The stability of the solution may be harmed if the surfactant concentration is too high because it may become overly viscous and form an impenetrable layer. The stability of the surfactant is also influenced by the pH of the solution. The majority of surfactants are most stable at a pH of 7. The stability of the solution can be decreased by surfactant molecule breakdown brought on by an excessively high or low pH. Surfactant stability is also influenced by the temperature of the solution, that is, lower temperatures can result in the surfactants precipitating out of solution, while higher temperatures typically increase stability. The stability of the surfactant may also be impacted by the presence of other compounds in the solution. The surfactant stability can be decreased by interactions with certain chemicals, such as salts. Making sure that all these elements in the solution are compatible with the surfactants is crucial for their stability. Some of them are given below:

1. Balancing between hydrophilic and hydrophobic molecules: the hydrophilic (attracted to water) head of the surfactant molecule and the hydrophobic (repelled by water) tail (not attracted to water). When the equilibrium between hydrophobic and hydrophilic molecules is preserved, a surfactant solution is stable.
2. Size of the head and tail: the solubility of the surfactant in a solution depends on the size of the head and tail. The surfactants with larger heads and tails are more soluble.
3. Charge: the charge of the surfactant molecule has the potential to impact stability. Positively charged surfactants tend to be more stable, whereas negatively charged surfactants are less stable.
4. Temperature: the stability of a surfactant solution can be impacted by temperature. Lower temperatures increase stability.
5. Solvent: the choice of solvent can have an impact on how stable a surfactant solution is. A surfactant's stability is more likely to eteriorate in polar solvents like water than in non-polar solvents like oils.
6. Critical micelle concentration (CMC): the CMC is the amount of surfactant molecules present in a solution that is required for the formation of micelles. Micelles are surfactant molecule aggregations that form a stable, spherical structure that can aid in stabilizing emulsions and raising surface tension.
7. pH: surfactant stability is significantly influenced by the pH of a solution. Surfactants are more soluble and stable at higher pH levels.
8. Additives: surfactant solutions can be supplemented with additives to improve the stability of the surfactant. Electrolytes, alcohols, and polymers are examples of typical additives.
9. Molecular weight: the stability of a surfactant in aqueous solutions is significantly influenced by its molecular weight. Surfactants with a higher molecular weight are typically more stable than those with a lower molecular weight.
10. Chain length: the chain length of a surfactant has a significant impact on its stability in aqueous solutions. Short-chain surfactants are frequently less stable than long-chain surfactants.

4.5 Lifetime of Surfactant Sensors

Depending on the type of sensor, the usage environment, and the frequency of use, the lifespan of surfactant sensors can vary significantly. Surfactant sensors typically survive up to 10 years, while some can last far longer, depending on the usage circumstances. The lifespan of a surfactant sensor may be impacted by the following factors:

1. Chemical exposure: some substances, such as acids and bases, which can corrode the electrodes, can harm sensors.
2. Temperature: extreme heat will short circuit a surfactant sensor by causing condensation to form on the electrodes; low temperatures will prolong the life of a surfactant sensor.
3. Humidity: excessive humidity can lead to electrode degradation, while low humidity can lead to the sensor drying up and malfunctioning.
4. Electrical interference: electrical interference can affect the sensor readings and lead to the sensor failing before its time.
5. Contamination: the sensor may malfunction or fail if it is contaminated with dust, dirt, oil, or other contaminants.
6. Use frequency: using a sensor frequently might shorten its lifespan by causing it to wear out more quickly than if it is used less frequently.

Surface bubbles are a common occurrence in many natural phenomena and technical applications. There are several theories and some experimental research, but a thorough comprehension of the lifetime phenomenon is still lacking. In view of the simulations and past findings, we will also analyze the trends in the lifetime of bubbles and drops, which have been documented in earlier experimental research.

The lifespan of surfactant sensors can be increased in the following ways:

1. Consistent maintenance: consistent maintenance is necessary to prolong the life of the sensors and ensure that they are functioning properly. Cleaning and part replacement should both be done on a regular basis.
2. Temperature control: keeping temperatures under control can extend the lifespan of surfactant sensors. Keep the temperature within the prescribed range and store the sensors in a cool, dry location.
3. Correct installation: surfactant sensors can be damaged and have their lifespan shortened if they are not installed properly. For the sensors to have the longest possible lifespan, it is crucial to make sure they are installed correctly.
4. Consistent calibration: the sensors must be calibrated consistently in order for them to function accurately and effectively. Their lifespan will be extended as a result of this.
5. Protective coating: to shield the sensors from the elements and lengthen their lifespan, a protective coating can be added.

4.6 Reuse of Surfactant Sensors

Many industries, such as cosmetics, laundry detergents, and cleaning goods, allow the reuse of surfactants. It can be done by employing a variety of methods, including recycling, centrifugation, distillation, and electrolysis. Used surfactants must be broken down into their component elements in order to be recycled. Surfactants and water or other contaminants are separated using centrifugation. Using evaporation, the process of distillation allows components of a mixture to be separated. An electric current can be used to purify compounds in a process called electrolysis. Reusing surfactants can cut waste and save money by lowering environmental pollution.

When reusing surfactants, however, the following safety measures must be taken:

1. Verify that no dangerous substances are present in the surfactants.
2. When handling surfactants, keep your skin and eyes away from contact, and use protective gear.
3. Ensure that surfactants are stored safely away from heat sources, sparks, and open flames.
4. In compliance with local laws, dispose of any unused or contaminated surfactant.
5. Before reusing a surfactant, familiarize yourself with its risks by consulting safety data sheets (SDSs).

4.7 Chemical Surfactants and Sensing

In order to simulate the function of olfaction, many chemical sensor arrays have been created with the intention of integrating an electronic nose with a range of materials. Figure 4.4 shows chemical surfactants in chemical sensing. Combining pattern recognition techniques with chemical sensor arrays replicates biological sensory recognition techniques. There are electronic nose systems for sale that are used in the food sector to ensure product quality. Current research focuses on the use of consumer-grade wearable electronic devices and commercial equipment to implement the electronic nose principle in environmental monitoring and medicine.

The idea that analytes will interact with various materials in different ways is at the core of chemical sensor arrays. As a result, any type of material may be utilized in sensors.[27] Polyacetylene, polythiophene, and polyaniline are some of the substances that can be chemically doped to make them conductors. The chemical and morphological properties of conductive polymers are affected by monomer identity, polymerization conditions, and device fabrication techniques.

Figure 4.5 shows the structure of surfactants and micelle formation. The ions that surround ionic surfactant micelles in solution, referred to as counterions, are attracted electrostatically. The effects of micelle charge on the structure of the surrounding solvent are visible at substantial distances from the micelle, despite the fact that the closest counterions can partially hide a charged micelle. The electrical conductivity of the mixture is one of the numerous qualities influenced by ionic micelles. Smaller ionic micelles

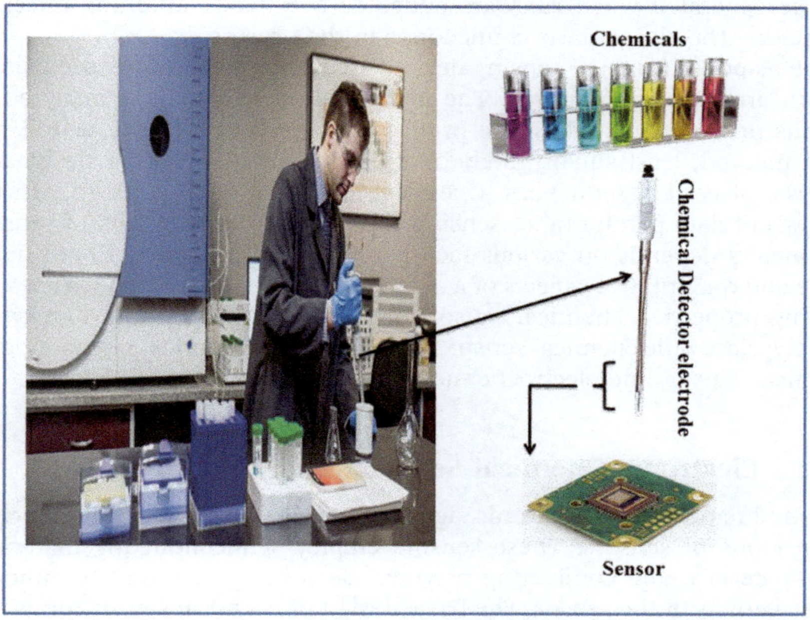

Figure 4.4 Usage of surfactants in chemical detection.

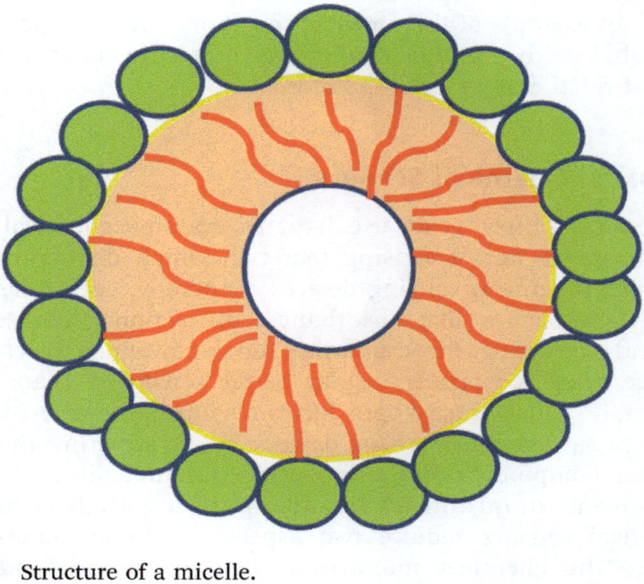

Figure 4.5 Structure of a micelle.

can be generated when salts are added to a colloid containing micelles, weakening the electrostatic connections in the process.

The response from the sensing device is recognized and processed using a known array of data patterns. The identification of signals is analyzed by various processes including the principal component method, neural network method, least square method, and machine learning for designated analysis of certain sequences. Chemical sensors play a major role in the analysis of data pertaining to sensing devices. The investigation of sensing parameters depends on various factors to give information on both qualitative and quantitative aspects of a sensor. Based on the techniques used for sensing properties, chemical sensors are usually classified into four types, namely, electronic chemical sensor, acoustical wave chemical sensor, optical chemical sensor, and electrochemical sensor.

4.7.1 Electronic Chemical Sensors

The modulation of an electronic signal has become of prime importance in these kinds of sensors. These sensors employ semiconducting materials, nanomaterials, and conducting polymers as sensing materials to enhance the sensitivity of the sensor. The basic design of chemical sensor comprises chemiresistors with capacitors and transistors in the architecture of a sensor. The variation in resistance is due to chemisorption or physisorption of the target molecules.[12] Various categories of materials for use of nanoelectronic detection systems have been discovered, notably graphene, carbon nanotubes, and 2D and 3D framework materials. Leading-edge location flaws and surface modification by covalent or non-covalent modifications are utilized as host–guest interaction sites in graphene-based materials, respectively.[30] An example of this involves the alteration of single-walled carbon nanotubes with multiple important protective factors to enable separation of volatile organic compounds.

4.7.2 Optical Chemical Sensors

Optical chemical sensor grids use light to examine chemical relations between an analyte and a sensing material, which distinguishes them from electronic chemical sensing devices. Generally, optical sensors use quantitative approaches like absorbance, diffraction, fluorescence, refraction, and scattering for examining how molecules react to light. Fluorescence sensors typically display better sensitivity than other imaging sensors.[31] The phrase "optoelectronic nose" has been utilized to categorize optical chemical sensor devices by constructing thumbprints for particular compounds and using predictive modelling to recognize those components in mixtures.[32] In addition to selectively engaging with solutes, optical sensors require two aspects: a probe material and a fluorophore. The chemical interactions between probes and analytes should be properly taken into account when developing cross-reactive optical and fluorescent matrices.[33,34]

4.7.3 Acoustic Wave Chemical Sensors

Acoustic wave devices, thickness shear mode resonators (TSM), and quartz crystal microbalances are just a few of the very broad categories of wave device technology. These devices oscillate at fixed frequencies, and variations in the mass of the device modulate particular frequencies. These devices can be modified using a number of the materials that have previously been described as also being suitable for chemical sensor arrays. The broad range of inter-molecular reactivity and specific interactions with such a variety of compounds that each of these materials exhibits, when combined, permits the fingerprint detection of individual compounds in mixtures.[35]

Wave devices can be adapted for various combinations of volatile organic chemicals, hydrogen gas, and mercury vapor. These materials use micro-fabricated metal-oxide cantilevers that have polymer sheets on top of them. These higher-order sensors use bulk and surface acoustic wave devices, in addition to electrical and optical signaling molecule channels.[11]

4.7.4 Electrochemical Sensors

Electrodes are different classes of material that can be utilized in electrochemical sensing devices. Electronic tongues are a common name for electrochemically based sensors. The surface of an electrode can be altered, which allows the system to target specific molecular interactions. The ability of partially permeable membrane materials to specifically oxidize or reduce targeted particles allows electrodes to be converted into sensors.[37] The ability to monitor the amounts of nitrate, nitrite, and ammonium in an aqueous phase has been demonstrated for a number of selectively permeable membrane sensors built from potentiometric polymers such as poly(vinyl chloride). The technique, which has been pioneered using these voltametric and potentiometric methods, is a popular subject of study since it provides classification of redox products as well as multi-analyte analysis of aqueous solutions such as cerebrospinal fluid.[36]

4.8 Surfactant Sensors in Chemical Detection and Application

4.8.1 Anionic Surfactant Sensors

Anionic surfactant sensors are used in thin films for analyzing conjugated polydiacetylenes based on imidazolium. In recent years, research on polydiacetylenes (PDAs), which are produced by UV irradiating self-assembled diacetylene (DA) supramolecules, has received a lot of attention.

Numerous PDA-based chemosensors have been developed as a result of its stimulus-induced blue-to-red shift. An example of an anionic surfactant-selective sensor based on PDA and an imidazolium polymer exhibited fluorescence amplification and colorimetric alterations in the presence of several anionic surfactants.[37] The conducting nature of poly(3,4-ethylenedioxythiophene) was significantly increased by incorporating anionic surfactants into the polymer solution, producing poly(styrenesulfonate) films. Addition of anionic surfactants to PEDOT:PSS aqueous solution has been suggested as a strategy to improve the conductivity of PEDOT:PSS film. According to the study, TsO anions improved the conductivity of PEDOT:PSS when added to its aqueous solution. The preparation of conducting polymer films involved spin-coating the polymer solution on glass substrates.[38]

Anionic sensors can be employed in cationic–anionic surfactant two-phase systems where temperature-sensitive aqueous surfactants manifest interesting phase behaviors and features. The cationic and anionic surfactant aqueous mixtures are mostly the result of electrostatic interactions between opposing charged head groups.[8,41] These interactions can be tuned to form different microstructures with distinctive geometries ranging from spherical to cylindrical to planar, which have been extensively researched both theoretically and empirically in the literature.[39]

4.8.2 Cationic Surfactant Sensors

A cationic surfactant–liquid crystal sensor platform for detecting thrombin has been developed. Based on the high specificity of aptamer identification, the liquid crystal (LC) sensing platform was shown to be effective. At first, LCs doped with octadecyl-trimethyl ammonium bromide (OTAB), a cationic surfactant, adopted a homeotropic orientation matching a dark optical picture. Thrombin and the matching aptamer were used to build the LC sensing platform.[40] The detection of thrombin by the LC sensing platform exhibited outstanding, consistent performance. A sensitive quinoline yellow sensor based on liquid crystals coated with cationic surfactants exhibited effective results in chemical detection. One of the synthetic colorings widely used in the food business is quinoline yellow (QY). The development of precise and straightforward QY detection procedures is crucial to guarantee food safety. In order to lower risk, it is crucial to effectively detect this food colorant. By using a cationic surfactant-decorated LC interface, an inventive liquid crystal (LC)-based sensor was created for the label-free and ultrasensitive detection of the QY.[41] This work is the first to describe how cetyltrimethylammonium bromide (CTAB) affects the electrochemical behavior of folic acid (FA) on the surface of a carbon paste electrode (CPE) in actual samples. Cyclic voltammetry (CV), chronocoulometry, and differential pulse voltammetry (DPV) were used to conduct the experiments.[45] Creation of a folic acid (FA) sensor based on the CTAB enhancing effect. Various electrochemical analyses of FA in the presence of CTAB have been explored.[42,46]

4.8.3 Non-ionic Surfactant Sensors

Poly ethoxylated non-ionic surfactant determination using a potentiometric sensor has been investigated. Different ethoxylated non-ionic surfactants (EONSs) and polyethylene glycols were examined using a novel liquid membrane surfactant sensor based on a 1,3-didecyl-2-methylimidazolium–tetraphenylborate ion-exchange complex. Amphipathic compounds known as non-ionic surfactants do not ionize in aqueous solution.[43] The potentiometric behavior of various barium-polyethoxylate complexes, including ethylene oxide units, has also been studied because ethoxylates have significant applications as surfactants.[44] The determination of the homologs of polyoxyethylated nonylphenols used arrays of non-selective non-ionic-surfactant sensors. It was shown that polyoxyethylated nonylphenols with various numbers of oxyethyl groups, which are homologs of non-ionic surfactants, may be independently determined using multisensor systems of the electronic tongue type based on non-selective solid-contact potentiometric sensors. A novel barium pseudocationic combination of a strongly ethoxylated fatty alcohol polyglycol ether and tetraphenylborate has been developed as the basis for a liquid membrane non-ionic surfactant-sensitive electrode. The compound has been integrated into a PVC membrane that has been plasticized and employed as a sensing material.[45]

4.8.4 Amphoteric (Zwitterionic) Surfactant Sensors

The effective reusability of a taste sensor with lipid polymer membranes *via* amphoteric surfactant cleaning has shown enhanced results in chemical detection. The purpose of the study was to improve the effectiveness of the cleaning procedure in order to make a taste sensor more reusable. This study used hydrophobic lipid polymer membranes in a surfactant cleaning approach for two different types of bitterness sensors. Four different types of surfactants with various polarities and ionicities were tested for usability in order to choose the best ones for cleaning. Use of both the normal cleaning solution and the LDA acid solution together significantly increased the reusability of an astringency sensor in numerous measurements of black tea. Effects of zwitterionic surface activity wormlike micelles driven by Na^+, ionic liquid, and anionic surfactant. The physicochemical characteristics of a mixed sodium dodecyl sulfate/*N*-alkyl-*N*′-carboxymethylimidazolium inner salt ionic liquid with zwitterionic surface activity has been investigated. The mixture under study displayed a stronger synergism, that is, more negative interaction characteristics, when compared to the mixed zwitterionic betaine surfactant/SDS system. Acetylene black electrodes enhanced with zwitterionic surfactant and carbon nanotubes to detect triclosan in household products have been examined.[11]

4.9 Biochemical Surfactants and Sensing

Figure 4.6 shows surfactants used in biochemical detection. Surfactants have emerged as versatile materials for many biological applications. The search

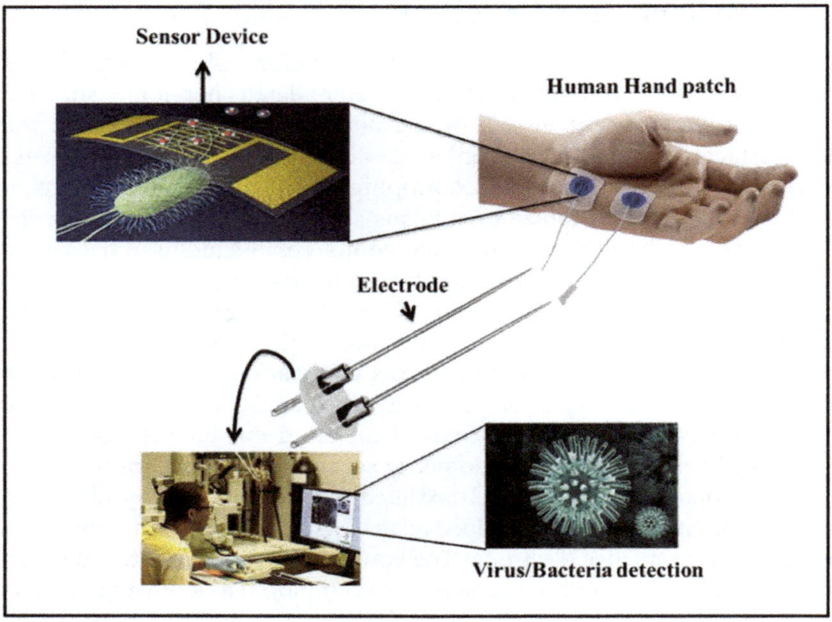

Figure 4.6 Usage of surfactants in biochemical detection.

for new surfactants for the betterment of biological applications is a continuous process, and in this regard bio-surfactants are considered to be favorable compounds. The use of bio-surfactants, bio-transformation and biodegradation has been discussed deliberately. The use of enzymes such as protease and amylase requires bio-technological skills for analysis of enzyme production, isolation, thermal behavior, storage stability, and pH stability. The invention of nanoparticle technology has profound applications in many biological activities with enhanced characteristic properties, such as enzyme encapsulation in spherulites, colloidal gas aphrons using multilayer surfactants, recovery of cells, proteins, biological molecules, and gas transformation in bio-reactors. The recent advancement in high-pressure phase transformation in phospholipid bilayers has become a model system for detection of biomembranes. Surfactant membranes are used in medicine, biotechnology, as an environmental protective shield for isolation of hydrocarbons and gases, and for the separation of metal ions, amino acids, and other suspensions. Sugar surfactants are used in the determination of various properties related to microemulsions.[8]

In order to lower surface tension at the air–liquid interface of the alveolus, pulmonary surfactant, a complex mixture of phospholipids and proteins, is used. The alveolus receives the generated, assembled, transported, and secreted surfactant, where it undergoes degradation and is subsequently recycled. In comparison to adults, babies, particularly premature ones, have a slower metabolism of surfactant. Respiratory distress from a problem with pulmonary surfactant metabolism is accompanied by morbidity and mortality. This happens because the metabolism of surfactants is impaired due to genetic abnormalities, proteolytic degradation, or rapid breakdown. The management of some of these conditions has been successful with the use of prenatal corticosteroids, surfactant replacement, whole lung lavage, and lung transplantation. Pulmonary surfactants lower surface tension at the alveolus' air–liquid contact, preventing the alveolus from collapsing during end-expiration. Additionally, it takes part in the host's natural defensive mechanism against inhaled infections.[50]

Non-ionic surfactants and organic solvents with low concentrations have enhanced the properties of cholesterol oxydase by increasing the activity of the enzyme. The cationic surfactants in reversed micelles in organic solvents emphasized two hydrolyzing enzyme proteases and amylase on detection of DNA from salmon testes. The evolution of surfactant-based sensors in biochemical detection is evidence of its various applications, which include medical and healthcare, antiviral activity, cyclic lipopeptide antibiotic, and clinical symptoms like irritation, skin dryness, itching, and stinging. In the view of all these, surfactant-based biochemical detection is very much necessary to prevent and to take measures after the serious effects. The most appropriate, simple, and cost-effective surfactant sensors are explored.

Electrochemical surfactant sensors will be described, focusing specifically on developments that have already been published.[51] In accordance with the actual measured value, electrochemical sensors are usually categorized as highly sensitive and selective, amperometric, conductometric, or impedimetric surface sensors.[5] Some of most significant are potentiometric surfactant sensors due to how simple and adaptable they are. There are various amperometric surfactant sensors that are real biosensors and involve living cells and microorganisms.

4.9.1 Potentiometric Surfactant Sensors

Even though several direct potentiometric sensors have been reported, potentiometric surfactant sensors are typically employed as quantitative terminal sensors. Among the most common types of sensor in this category are coated wire-type, liquid membrane, and ISFET (ion-selective field effect transistor) sensors. The sensors can be used to recognize non-ionic as well as ionic (cationic and anionic) surfactants.[6] They can be operated constantly in flow injection analysis (FIA) mode or intermittently in batch mode. Although several direct potentiometric sensors have been described, potentiometric surfactant sensors are typically employed as titration endpoint sensors.[47–49] For sensors designed using liquid membrane surfactants and coated wire it is indeed difficult to distinguish between liquid membrane sensors and coated wire sensors. These variants employ ISE membranes to operate. Since these membranes are generally solid for practical purposes, the term liquid membrane can be misleading. Yet, the smart phone complexing agents known as sites are typically included in a matrix gel after being dissolved in a suitable solvent.[50] An ion-selective electrode has been described as the end-point titration sensor for determining the concentration of anionic and cationic surfactants *via* titration. It demonstrated that surfactant mixtures and non-ionic surfactants can both be tested using the same electrode. As precisely and easily as the pure surfactant, the concentrations of combinations of anionic and non-ionic surfactants may be calculated.[51]

4.9.2 Impedimetric Surfactant Sensors

Direct affinity sensors have been presented with simple preparation approaches. The oligonucleotide or antibodies were similarly immobilized as a result of the self-consistent introduction of a hydrocarbon chain attached to an oligonucleotide pentadecathymidylate into the hydrophobic region of the surfactant bilayers or as a result of the adsorption of antibodies on the surface of the bilayers. Impedance spectroscopy was utilized to find the responses and antibody–antigen interactions increased the impedance, whereas particular DNA-coupling decreased the actual fraction of the impedance.[52] Easy procedures have been suggested for making direct affinity sensors. The generation of immunosensors is a result of antibodies adhering to bilayer surfaces. Significant alterations in impedance spectra have been seen in both trials. The results are encouraging for the development of impedimetric affinity sensors for use in healthcare or the environment. The lubricating qualities of motor oil can be adversely affected by as little as a few hundred ppm of water. The oil additives (surfactants) and water molecules started interacting right away after the injection, creating an inverse micelle-based water-in-oil emulsion. After emulsification, water in the solution was gradually removed by evaporation and electrolysis. A computer analysis model was created using the literature equations defining these interactions in order to support the experimental findings and demonstrate the kinetics of the water–oil interactions. The model enhanced understanding of the experimental data by illuminating the processes taking place in the water–oil system.

4.9.3 Conductometric Surfactant Sensors

This sort of surfactant sensor for conducting conductivity measurements to track the evolution of a number of particles in the emulsion polymerization has been discussed. The three main phases of the polymerizing latex were assumed to be equally divided among the surfactant in the developed model. A soft-sensor approach was then suggested for counting the amount of polymer particles by fusing the conductivity model with the available conversion, temperature, and conductivity signals (N-P). The major objectives were to watch the evolution of N-P (nucleation/coagulation) in real time under varied reaction scenarios and to validate the conductivity model over a larger variety of operational settings. Several batch and semi-batch polymerization operations were conducted.[53]

4.9.4 Amperometric Surfactant Sensors

Several real surfactant biosensors using microorganisms or cells are available alongside amperometric surfactant sensors. Numerous modified PVC membranes have been characterized based on their permeability and selectivity for electrochemically active substances. The species were identified using a two-electrode amperometric cell in phosphate buffer solution. The diffusing species were observed to be paracetamol, catechol, ascorbic acid, and hydrogen peroxide. Cationic and non-ionic surfactants were employed as the membrane modifying agents in the research.[54] The detection of anionic surfactants was achieved by building a *Pseudomonas rathonis* T-based amperometric biosensor. Plasmids used in surfactant degradation were found in microorganisms. Sodium dodecyl sulfate and volgonat were quite corrosive to the sensor. The selectivity of PVC membranes plasticized with surfactants and liquid crystals was studied using an amperometric sensor, and the results were presented. The effects of non-ionic surfactants on the electrochemical behavior of the compounds might be studied by including ubiquinone and menaquinone in a carbon paste electrode. Additionally, impacts on the sensitivity of the glucose biosensor were shown when a similar modification was applied to create an enzyme electrode in which the quinone molecule functions as a redox mediator.[55]

4.10 Surfactant Sensors in Biochemical Detection and Applications

4.10.1 Anionic Surfactant Sensors

A microbial sensor for the measurement of anionic surfactants has been investigated. For the purpose of detecting anionic surfactants in an aquatic environment a unique biosensor has been built. Analyzing anionic surfactants was quick and easy, and no hazardous organic reagents were needed.[56] Patients may get anaphylactic shock as a result of ingesting food allergens, particularly buckwheat proteins.[63] There is a need for the creation of an easy and quick screening approach using a field effect transistor biosensor for food allergens in food products. In this investigation, a buckwheat allergenic protein was successfully detected using FET.[57] Sodium dodecyl sulfide (SDS), an anionic surfactant, served as a signal amplifier for FET detection. For the purpose of detecting allergens in industrial domain, the suggested approach attained the desired level of sensitivity. An anionic surfactant-based electrochemical nicotine sensor based on cerium nanoparticles has been explored. The primary alkaloid in tobacco leaves (*Nicotianata bacum* L. Solanaceae) is nicotine, also known as

3-(1-methyl-2-pyrrolidinyl) pyridine. These leaves are more famous for their use in cigarettes than for their therapeutic uses.[17]

4.10.2 Cationic Surfactant Sensors

A glucose-modified carbon paste sensor in the presence of a cationic surfactant for mefenamic acid detection in urine and pharmaceutical samples has been studied. A unique and reliable carbon paste electrode (CPE) modified with glucose was created for the ultrasensitive detection of mefenamic acid (MA).[58,64] Diffusion–adsorption control was visible in the sensing results. The technique displayed a low detection limit and a broader linear dynamic range. Both biological and pharmacological samples of MA were successfully detected by the newly designed sensor.[58] A dangerous analog of glucose called alloxan (AL) selectively kills the pancreatic cells that make insulin, acting as a strong inducer of diabetes. In order to analyze AL in the presence of anthrone, a sensitive and focused carbon paste electrode immobilized with cetyltrimethylammonium bromide (CTAB) was used.[59]

A study of its increased voltammetric determination in cationic surfactant medium employing a boron-doped diamond electrode served as the first electrochemical evaluation of the antiviral drug favipiravir used to treat COVID-19.[60] The lipophilicity affects the membrane resistance, ionophore solubility, ion exchange across the membrane, and ultimately the analytical signal. All of the examined cationic surfactants could be titrated using potentiometric titration, and P1 plasticizer demonstrated exceptional titration performance, rapid reaction time, high signal change, and signal stability.[48]

4.10.3 Non-ionic Surfactant Sensors

A sensitive and specific protein detection method using a non-ionic surfactant-decorated liquid crystal sensor has been developed. In the human body, proteins are in charge of most biochemical processes. The creation of sensitive and focused protein detection techniques is crucial. This study used a liquid crystal-based sensor for highly sensitive and specific lysozyme detection. By employing an LC interface and dodecyl β-D-glucopyranoside, a non-ionic surfactant, concanavalin A and bovine serum albumin were created. *Pseudomonas aeruginosa* proliferation has been studied using a new sensor technique. To investigate how surfactants affect *Pseudomonas aeruginosa* growth in media, a novel sensor technology based on series piezoelectric quartz crystal sensing was developed. The frequency shift curves produced under various growth circumstances were compared with one another. Three kinetic parameters were obtained to explain the growth of microorganisms by fitting frequency shift curves. When it came to bacterial growth, surfactants were a growth inhibitor.[17]

Applications for mesostructured polyaniline in H_2O_2 and glucose sensing have been found utilizing a combination of mixed surfactants, anionic sodium dodecylsulfate, and non-ionic polymers. Mesostructured polyaniline was made by allowing a mixture of a non-ionic polymeric surfactant and sodium dodecyl sulfate, an anionic surfactant, to self-assemble. The approach used in this study created a novel platform for the creation of high-performance glucose and other biosensors based on polyanilines. Immobilizing the sensing material within the membrane with graphene nanoparticles lowered the electrical resistance and signal noise of the membrane by preventing the leaching of electroactive material from the membrane.[61,62]

4.10.4 Amphoteric (Zwitterionic) Surfactant Sensors

A biosensor based on laccase-halloysite nanotubes and imidazoliumzwitterionic surfactant for determining dopamine has been developed. Improved catalytic activity and stabilities in biosensing are provided by zwitterionic surfactants that integrate nanomaterials and biomolecule architectures.[9] A novel electroanalytical approach for the measurement of dopamine by square-wave voltammetry was created using the biosensor. This review summarizes the most recent developments in membranes, micro-nanocapsules, nano-microgels, as well as multilayered thin films with amphoteric properties. In light of their prospective use in nanotechnology, biotechnology, and medicine, as well as their stimuli responsive behavior, synthetic methods for amphoteric materials and physicochemical characteristics have been described.[51] As a result of their strong propensity for self-organization in aqueous conditions, amphoteric macromolecules with hydrophobic moieties can act as a crude model of biological membranes. For peroxidase immobilization in the creation of biosensors, gold nanoparticles have been disseminated in zwitterionic surfactants.

4.11 Conclusion

This chapter consists of basic information on the stability, life time, and reuse of surfactants, and their applications in chemical and biochemical detection. Here, the impact of surfactants in daily life has been discussed extensively along with their applications. Most of the surfactants are similar in nature, including a hydrocarbon chain which can be framed as linear, branched, or aromatic in nature, which enhances the hydrophilic property of the surfactants. Nonionic surfactants contain both hydrophilic and lipophilic groups. Surfactants are classified into four types, based on the charge of their polar head group: anionic, non-ionic, amphoteric and cationic surfactants. The creation, optimization, and proper dosage of surfactants are carried out through analysis of the interface. We can assess the effectiveness and efficiency of surfactants using tensiometers, which measure surface tension. They aid in the construction and optimization of emulsions by precisely measuring the interfacial tension. Current research focuses on the use of consumer-grade wearable electronic devices and commercial equipment to implement the electronic nose principle in environmental monitoring and medicine. The idea that analytes will interact with various materials in different ways is at the core of chemical sensor arrays. As a result, any type of material may be utilized in sensors. The use of bio-surfactants, bio-transformation and bio-degradation has been reviewed in this chapter. The use of enzymes such as protease and amylase requires bio-technological skills for analysis of enzyme production, isolation, thermal behavior, storage stability, and pH stability. The invention of nano-technology has profound applications in many biological activities with enhanced characteristic properties.

Acknowledgements

One of the authors acknowledges the management of Ballari Institute of Technology and Management, Ballari for their support.

References

1. A. V. Kim, N. N. Medvedev and A. Geiger, Molecular dynamics study of the volumetric and hydrophobic properties of the amphiphilic molecule, *J. Mol. Liq.*, 2014, **189**, 74.
2. M. Mobin, R. Aslam and J. Aslam, Non toxic biodegradable cationic gemini surfactants as novel corrosion inhibitor for mild steel in hydrochloric acid medium and synergistic effect of sodium salicylate: Experimental and theoretical approach, *Mater. Chem. Phys.*, 2017, **191**, 151.
3. J. D. York and A. Firoozabadi, Comparing effectiveness of rhamnolipid biosurfactant with a quaternary ammonium salt surfactant for hydrate anti-agglomeration, *J. Phys. Chem. B*, 2008, **112**(3), 845.
4. C. Cowan-Ellsberry, S. Belanger, P. Dorn, S. Dyer, D. Mcavoy and H. Sanderson, Environmental safety of the use of major surfactant classes in North America, *Crit. Rev. Environ. Sci. Technol.*, 2014, **44**(17), 1893.
5. J. Ju and W. Chen, In situ growth of surfactant-free gold nanoparticles on nitrogen-doped graphene quantum dots for electrochemical detection of hydrogen peroxide in biological environments, *Anal. Chem.*, 2015, **87**(3), 1903.
6. G. Ziyatdinova, E. Ziganshina and H. Budnikov, Voltammetric sensing and quantification of eugenol using nonionic surfactant self-organized media, *Anal. Methods*, 2013, **5**(18), 4750.
7. A. A. P. Khan, A. Khan, M. M. Rahman and A. M. Asiri, Conventional surfactant-doped poly(o-anisidine)/GO nanocomposites for benzaldehyde chemical sensor development, *J. Sol–Gel Sci. Technol.*, 2016, **77**(2), 361.
8. G. Tigari and J. G. Manjunatha, Optimized Voltammetric Experiment for the Determination of Phloroglucinol at Surfactant Modified Carbon Nanotube Paste Electrode, *Instrum. Exp. Tech.*, 2020, **63**(5), 750.

9. O. Galovic, M. Samardzic, M. Hajdukovic and M. Sak-Bosnar, A new graphene-based surfactant sensor for the determination of anionic surfactants in real samples, *Sens. Actuators, B*, 2016, **236**, 257.
10. D. Madunic-cacic, M. Sak-Bosnar, O. Galovic, N. Sakac and R. Matesic-Puac, Determination of cationic surfactants in pharmaceutical disinfectants using a new sensitive potentiometric sensor, *Talanta*, 2008, **76**(2), 259.
11. N. Sakac, D. Markovic, B. Šarkanj, D. Madunic-Cacic, K. Hajdek and B. Smoljan, Direct potentiometric study of cationic and nonionic surfactants in disinfectants and personal care products by new surfactant sensor based on 1,3-dihexadecyl-1*H*-benzo[*d*]imidazol-3-ium, *Molecules*, 2021, **26**(5), 1.
12. H. Li, Z. Deng, Q. Tian, L. Lun, P. Zhao and X. Yang, Application of carbon nanotubes and zwitterionic surfactant-modified acetylene black electrode for the determination of triclosan in household commodities, *Int. J. Environ. Anal. Chem.*, 2022, **102**(4), 987.
13. S. Gold, F. J. Ibanez and F. P. Zamborini, Chemiresistive Sensing of Volatile Organic Compounds with Films of Surfactant-Stabilized Gold and Gold Silver Alloy Nanoparticles, *ACS Nano*, 2008, **2**(8), 1543.
14. A. Pastore, D. Badocco and P. Pastore, Talanta Influence of surfactant chain length, counterion and OrMoSil precursors on reversibility and working interval of pH colorimetric sensors, *Talanta*, 2020, **212**, 120739.
15. D. Jadreško and M. Zelić, Square-wave voltammetry of electroinactive surfactants, *Electroanalysis*, 2015, **27**, 1669.
16. S. Sharvelle, R. Lattyak and M. K. Banks, Evaluation of biodegradability and biodegradation kinetics for anionic, nonionic, and amphoteric surfactants, *Water, Air, Soil Pollut.*, 2007, **183**(1–4), 177.
17. A. J. Frend, G. J. Moody, J. D. R. Thomas and B. J. Birch, Flow injection analysis with tubular membrane ion-selective electrodes in the presence of anionic surfactants, *Analyst*, 1983, **108**, 1357.
18. N. Sakač, M. Jozanović, M. Karnaš and M. Sak-Bosnar, New Sensor for Determination of Anionic Surfactants in Detergent Products with Carbon Nanotubes as Solid Contact, *J. Surfactants Deterg.*, 2017, **20**, 881.
19. M. Hajdukovic, M. Samardzic, O. Galovic, A. Szechenyi and M. A. Sak-Bosna, functionalized nanomaterial based, new, solid state cationic-surfactant-selective sensor with fast response and low noise, *Sens. Actuators, B*, 2017, **251**, 795.
20. H. Harma, S. Laakso, S. Pihlasalo and P. Hanninen, A Fluorometric fullerenol sensor for rapid detection of ionic and non-ionic surfactants, *Tenside, Surfactants, Deterg.*, 2010, **47**(1), 40.
21. R. A. Salkar, P. Jeevanandam, G. Kataby, S. T. Aruna, Y. Koltypin and O. Palchik, Elongated Copper Nanoparticles Coated with a Zwitterionic Surfactant, *J. Phys. Chem. B*, 2000, **104**(5), 893.

22. Y. Tian, R. G. Holt and R. E. Apfel, Investigation of liquid surface rheology of surfactant solutions by droplet shape oscillations: Experiments, *J. Colloid Interface Sci.*, 1997, **187**(1), 1.
23. J. Eastoe and J. S. Dalton, Dynamic surface tension and adsorption mechanisms of surfactants at the air–water interface, *Adv. Colloid Interface Sci.*, 2000, **85**(2), 103.
24. A. P. Santos and A. Z. Panagiotopoulos, Determination of the critical micelle concentration in simulations of surfactant systems, *J. Chem. Phys.*, 2016, **144**(4), 044709.
25. D. Rana, G. H. Neale and V. Hornof, Surface tension of mixed surfactant systems: Lignosulfonate and sodium dodecyl sulfate, *Colloid Polym. Sci.*, 2002, **280**(8), 775.
26. D. R. Otis Jr, E. P. Ingnetio, R. D. Kamm and M. Johnson, Dynamic surface tension of surfactant TA: Experiments and theory, *J. Appl. Physiol.*, 1994, **77**(6), 2681–2688.
27. K. Babu, N. Pal, A. Bera, V. K. Saxena and A. Mandal, Studies on interfacial tension and contact angle of synthesized surfactant and polymeric from castor oil for enhanced oil recovery, *Appl. Surf. Sci.*, 2015, **353**, 1126.
28. N. D. Denkov, S. Tcholakova, K. Golemanov and K. P. Ananthpadmanabhan, The role of surfactant type and bubble surface mobility in foam rheology, *Soft Matter*, 2009, **5**(18), 3389–3408.
29. A. Bhattacharyya, F. Monroy, D. Langevin and J. F. Argillier, Surface rheology and foam stability of mixed surfactant-polyelectrolyte solutions, *Langmuir*, 2000, **16**(23), 8727.
30. N. Abramova, Application of Photocured Polymer Ion Selective Membranes for Solid-State Chemical Sensors, *Chemosensors*, 2015, **3**, 190.
31. O. J. DSouza, R. J. Mascarenhas, T. Thomas, B. M. Basavaraja, A. K. Saxena and K. Mukhopadhyay, Platinum decorated multiwalled carbon nanotubes/Triton X-100 modified carbon paste electrode for the sensitive amperometric determination of paracetamol, *J. Electroanal. Chem.*, 2015, **739**, 49.
32. A. K. Dwivedi, M. Pandeeswar and T. Govindaraju, Assembly modulation of PDI derivative as a supramolecular fluorescence switching probe for detection of cationic surfactant and metal ions in aqueous media, *ACS Appl. Mater. Interfaces*, 2014, **6**(23), 21369.
33. R. Kumar and S. R. Raghavan, Photogelling fluids based on light-activated growth of zwitterionic wormlike micelles, *Soft Matter*, 2009, **5**(4), 797.
34. L. Ding, S. Wang, Y. Liu, J. Cao and Y. Fang, Bispyrene/surfactant assemblies as fluorescent sensor platform: Detection and identification of Cu^{2+} and Co^{2+} in aqueous solution, *J. Mater. Chem. A*, 2013, **1**(31), 8866.
35. I. Thivaios, V. Koukoumtzis, J. K. Kallitsis and G. Bokias, Quinoline-labeled poly(N-isopropylacrylamide): A selective polymeric luminescent sensor of cationic surfactants, *Sens. Actuators, B*, 2016, **233**, 127.

36. F. Li, J. H. C. Wang and Q. M. Wang, Thickness shear mode acoustic wave sensors for characterizing the viscoelastic properties of cell monolayer, *Sens. Actuators, B*, 2008, **128**(2), 399.
37. M. M. Charithra and J. G. Manjunatha, Enhanced voltammetric detection of paracetamol by using carbon nanotube modified electrode as an electrochemical sensor, *J. Electrochem. Sci. Eng.*, 2020, **10**(1), 29.
38. V. B. Patil, S. J. Malode, S. M. Tuwar and N. P. Shetti, Graphene sheet-based electrochemical sensor with cationic surfactant for sensitive detection of atorvastatin, *Sens. Int.*, 2022, **3**, 100198.
39. S. Lee, K. M. Lee, M. Lee and J. Yoon, Polydiacetylenes bearing boronic acid groups as colorimetric and fluorescence sensors for cationic surfactants, *ACS Appl. Mater. Interfaces*, 2013, **5**(11), 4521.
40. A. Pasha, S. Khasim, O. A. Al-Hartomy, M. Lakshmi and K. G. Manjunatha, Highly sensitive ethylene glycol-doped PEDOT-PSS organic thin films for LPG sensing, *RSC Adv.*, 2018, **8**(32), 18074.
41. P. A. Pushpanjali, J. G. Manjunatha and M. T. Shreenivas, The Electrochemical Resolution of Ciprofloxacin, Riboflavin and Estriol Using Anionic Surfactant and Polymer-Modified Carbon Paste Electrode, *Anal. Chem.*, 2019, **4**, 13427.
42. A. Sein and J. B. F. Engberts, Micelle to Lamellar Aggregate Transiiion of an Anionic Surfactant in Dilute Aqueous Solution Induced by Alkali Metal Chloride and Tetraalkylammonium Chloride Salts, *Langmuir*, 1995, **11**(2), 455.
43. Y. Wang, Q. Hu, T. Tian, Y. Gao and L. Yu, A nonionic surfactant-decorated liquid crystal sensor for sensitive and selective detection of proteins, *Anal. Chim. Acta*, 2016, **937**, 119.
44. F. S. Mohseni-Shahri, F. Moeinpour and A. Verdian, A cationic surfactant-decorated liquid crystal-based sensor for sensitive detection of quinoline yellow, *Sci. Rep.*, 2021, **11**(1), 1.
45. N. S. Prinith and J. G. Manjunatha, Materials Science for Energy Technologies Surfactant modified electrochemical sensor for determination of Anthrone – A cyclic voltammetry, *Mater. Sci. Energy Technol.*, 2019, **2**(3), 408.
46. M. Behpour, A. M. Attaran, M. M. Sadiany and A. Khoobi, Adsorption effect of a cationic surfactant at carbon paste electrode as a sensitive sensor for study and detection of folic acid, *Measurement*, 2016, **77**, 257.
47. M. Sak-Bosnar, M. Samardzic and O. Galovic, The influence of ethoxylated nonionic surfactants on the potentiometric determination of anionic surfactants, *Int. J. Electrochem. Sci.*, 2011, **6**(3), 561.
48. M. Sak-Bosnar, D. Madunic-Cacic, N. Sakac, O. Galovic, M. Samardzic and Z. Grabaric, Potentiometric sensor for polyethoxylated nonionic surfactant determination, *Electrochim. Acta*, 2009, **55**(2), 528.
49. N. M. Mikhaleva and E. G. Kulapina, Arrays of nonselective nonionic-surfactant sensors for the separate determination of the homologues of polyoxyethylated nonylphenols, *J. Anal. Chem.*, 2005, **60**(6), 646.

50. N. Hareesha and J. G. Manjunatha, Surfactant and polymer layered carbon composite electrochemical sensor for the analysis of estriol with ciprofloxacin, *Mater. Res. Innovations*, 2019, **24**(6), 349–362.
51. C. Raril and J. G. Manjunatha, A simple approach for the electrochemical determination of vanillin at ionic surfactant modified graphene paste electrode, *Microchem. J.*, 2020, **154**, 104575.
52. N. Sakac, D. Markovic, B. Sarkanj, C. Madunic, D. Acic, K. Hajdek, B. Smoljan and M. J. Direct, Potentiometric Study of Cationic and Nonionic Surfactants in Disinfectants and Personal Care Products by New Surfactant Sensor Based on 1,3-Dihexadecyl 1*H*-benzo[*d*]imidazol 3-ium, *Molecules*, 2021, **26**, 1366.
53. O. Fizer, M. Fizer, V. Sidey and Y. Studenyak, Predicting the end point potential break values: A case of potentiometric titration of lipophilic anions with cetylpyridinium chloride, *Microchem. J.*, 2021, **160**, 105758.
54. M. Karnas, N. Sakac, M. Jozanovic, M. Tsakiri, M. Kopriva, E. K. Andrić and M. Sak-Bosnar, The Influence of Plasticisers on Response Characteristics of Anionic Surfactant Potentiometric Sensor, *Int. J. Electrochem. Sci.*, 2017, **12**, 5921.
55. M. Jozanovic, N. Sakac, M. Karnas and M. Medvidovic-Kosanovic, Potentiometric Sensors for the Determination of Anionic Surfactants–A Review, *Crit. Rev. Anal. Chem.*, 2021, **51**(2), 115.
56. J. Sanchez and M. Del Valle, Determination of anionic surfactants employing potentiometric sensors – A review, *Crit. Rev. Anal. Chem.*, 2005, **35**(1), 15.
57. D. Madunic-Cacic, M. Sak-Bosnar, R. Matesic-Puac and M. Samardzic, Potentiometric determination of anionic surfactants in formulations containing cocoamidopropyl betaine, *Int. J. Electrochem. Sci.*, 2012, **7**(1), 875.
58. S. Duval, M. Keddam, M. Sfaira, A. Srhiri and H. Takenouti, Electrochemical Impedance Spectroscopy of Epoxy-Vinyl Coating in Aqueous Medium Analyzed by Dipolar Relaxation of Polymer, *J. Electrochem. Soc.*, 2002, **149**(11), 520.
59. H. M. Naien, S. Dzyadevych, F. Lagarde and N. Jaffrezic-Renault, A Conductometric Sensor Specific for Cationic Surfactants, *Electroanalysis*, 2012, **24**(6), 1441.
60. T. Thomas, R. J. Mascarenhas, O. J. D' Souza, S. Detriche, Z. Mekhalif and P. Martis, Pristine multi-walled carbon nanotubes/SDS modified carbon paste electrode as an amperometric sensor for epinephrine, *Talanta*, 2014, **125**, 352.
61. L. Taranova, I. Semenchuk, T. Manolov, P. Iliasov and A. Reshetilov, Bacteria-degraders as the base of an amperometric biosensor for detection of anionic surfactants, *Biosens. Bioelectron.*, 2002, **17**(8), 635.
62. M. Gao, L. Wang, J. Chen, S. Li, G. Lu and L. Wang, Aggregation-Induced Emission Active Probe for Light-Up Detection of Anionic Surfactants and Wash-Free Bacterial Imaging, *Chem. – Eur. J.*, 2016, **22**(15), 5107.

63. J. G. Manjunatha, A novel voltammetric method for the enhanced detection of the food additive tartrazine using an electrochemical sensor, *Heliyon*, 2018, **4**, e00986.
64. A. B. Monnappa, J. G. Manjunatha and A. S. Bhatt, Design of a sensitive and selective voltammetric sensor based on a cationic surfactant-modified carbon paste electrode for the determination of alloxan, *ACS Omega*, 2020, **5**(36), 23481.

CHAPTER 5

Surfactant-based Electrochemical Sensing and Bio-sensing: Safety and Toxicity

SHRIDEVI SALAGARE,*[a] S. ASHOKA[b] AND PRASHANTH S. ADARAKATTI*[c]

[a] Department of Chemistry and Research Centre, NMKRV College for Women, Bengaluru, India; [b] Department of Chemistry, School of Applied Science, REVA University, Bengaluru – 560064, India; [c] Department of Chemistry, SVM Arts, Science and Commerce College, ILKAL – 587125, affiliated to Rani Channamma University, Belagavi, Karnataka, India
*Emails: salagareshridevi31@gmail.com; prashanthsa143@gmail.com

5.1 Introduction

Surfactants are organic compounds composed of two chemical parts with different polarities, a head group with affinity for polar phases and a tail group that is attracted to nonpolar phases.[1–7] Due to their structural uniqueness, surfactants can be widely used to reduce the surface and interfacial tension between two or more phases. Their tendency to generate self-assembled structures in solution can also lead to the formation of micelles with diameters ranging from nanometers to microns.[8–13] The amphiphilic nature of surfactants makes them suitable for use in numerous industrial products, including medicines,[14–16] corrosion inhibitors for protecting steel and other corrosive metals,[17–19] detergents,[20–22] de-emulsifiers,[23–25] wetting agents,[26,27] oil recovery enhancers,[28–31]

Detection Science Series No. 23
Surfactant-based Sensors in Chemical and Biochemical Detection
Edited by J. G. Manjunatha

Published by the Royal Society of Chemistry, www.rsc.org

pour-point depressants,[32–34] pharmaceutical formulations,[35–37] drug delivery, and detection of heavy metal ions.[38–40]

In addition to industrial applications, surfactants have great potential to overcome the current limitations in nanotechnology.[41–46] For example, amphiphilic surfactants have been reported as stabilizers for the preparation of stable dispersions of hydrophobic inorganic nanomaterials such as carbon nanotubes,[47–49] graphene,[50] transition metal dichalcogenides,[51] and black phosphorus.[52,53] Since the first graphene exfoliation method, micromechanical exfoliation, was discovered in 2004, researchers have attempted scale-up of the process by stabilizing exfoliated graphene in solvents.[48] Considering the accessibility and environmental toxicity of organic solvents, water may be an ideal solvent system for preparing nanomaterial dispersions, but most inorganic nanomaterials have hydrophobic surfaces and thus exhibit repellent behavior in water. To overcome this limitation, various organic solvents have been exploited for stabilizing nanomaterials, such as *N*-methyl-pyrrolidone and dimethylformamide.[54–58] Although these organic solvents can successfully disperse nanomaterials, their high boiling points lead to the requirement for additional high-temperature annealing to remove the solvent, which may induce chemical degradation of the nanomaterials and obscure their toxicity limits for use in biological applications.[48] This hurdle was overcome by adding amphiphilic surfactants to water, together with inorganic nanomaterials. The hydrophobic parts of the surfactants interact with the surfaces of the nanomaterials, and the hydrophilic parts stabilize the surfactant-nanomaterial composite in water. Furthermore, the buoyant densities of nanomaterials can be controlled based on the type of surfactants, which enables the separation of nanomaterials in the structure *via* isopycnic density gradient ultracentrifugation.[59] The resulting structurally sorted nanomaterials have contributed significantly to many research fields in nanotechnology, especially electronics,[60–62] optoelectronics,[57,63,64] energy,[58,65–69] biology,[48] and catalysis.[54]

Due to the potential advantage of surfactants for tuning the surface/interface tensions between solid/liquid interfaces and further improving the dispersion stability, surfactants may be a key agent in the synthesis of nanoparticles with well-controlled geometries. Recently, researchers have focused on nanoparticle-based sensors that detect organic and inorganic pollutants and biomolecular compounds in serum or urine samples as biomarkers for the early diagnosis of diseases. One approach is to use localized surface plasmon resonance phenomena on a nanoparticle array, where the nanoparticle size and shape, degree of aggregation, and refractive index affect the sensitivity of the sensors. Furthermore, advanced sensing systems have been developed for enhancing both the identification and transduction processes by using many novel nanoparticles rather than conventional organic dye-based sensing assays, where the former offer advantages in sensitivity, selectivity, stability, reusability, and practical applicability. Among the available methods, electrochemical and biosensor methods are simple, highly efficient, and exhibit great potential for point-of-care diagnosis because the detection process is facile, as responses can be discerned by the naked eye or *via* simple instrumental techniques. For electrochemical and biosensor applications, the unique physical and chemical properties (size, shape, large active surface area, and higher activity) of nanoparticles afford tremendous catalytic activities that differ from those of their bulk counterparts due to size-induced quantum confinement effects in the former. For example, colloidal plasmonic gold and silver nanoparticles have been most intensively studied as they produce rapid and vivid responses to incident light and can be efficiently used for the sensitive detection of target analytes through visible color changes in the solution phase.

Surfactants have been widely used in the preparation of nanoparticles because they are electron-rich and provide good chelating centers for improving the stability of nanoparticles synthesized in a solution. One of the common surfactants for the synthesis of metallic nanoparticles is a gemini surfactant, which consists of two hydrophobic tails and two hydrophilic heads connected by a spacer.[70–74] Gemini surfactants are characterized by their relatively low critical micelle concentrations and high efficiency in decreasing the surface and interfacial tensions of water compared to conventional monomeric surfactants.[75–77] These properties provide a high adsorption affinity toward the surface of metal nanoparticles for high stabilization.

In this chapter, we provide an overview of surfactants for nanoparticle synthesis and discuss the detailed relationships between the molecular structure of the surfactant and the resulting particle structure distributions for optimizing ultrasensitive detection in sensing application devices.

5.2 Surfactant Structure-dependent Synthesis and Properties

5.2.1 Surfactant Modification for Nanoparticle Synthesis

The amphiphilic characteristics of surfactants play an important role in nanoparticle synthesis by preventing aggregation of the synthesized nanoparticles and providing more stability in colloidal systems.[43] Surfactant-assisted nanoparticle synthesis has attracted considerable attention in the development of nanoscale sensors for detecting toxic materials, biomaterials, and biomarkers. Because nanoscale sensors should distinguish a trace amount of material, it is necessary to achieve high sensitivity and selectivity for a specific analyte. The sensitivity and selectivity are strongly related to the structural parameters of the nanoparticles, including the precise size distribution. Consequently, controlling the size by modifying the molecular structure of the surfactant can provide a new strategy for maximizing the sensitivity enhancement.[78–82]

One such approach in which the tail length of the surfactant was altered to control the nanoparticle size, growth, and colloidal stability was recently reported and an electro-deposited graphene–Au-modified glassy carbon electrode with amplified signal as shown in Figure 5.1. The colloidal stability of surfactant–nanoparticle hybrid systems can be described by the thermodynamic Gibbs free energy changes ($\Delta G°$) and to understand the impact of the molecular structure of the surfactant on the stability of the synthesized nanoparticles in the colloidal state, we must first understand the behavior of surfactants in solution systems. Surfactants consist of two parts with opposing polarities, which makes the surfactant solution thermodynamically unstable due to the repulsive force between the hydrophobic tail of the surfactant and the polar medium in which the surfactant is dissolved. This repulsion forces the surfactant to be adsorbed on the nanoparticle and undergo micellization to stabilize the system by lowering the free energy. In the surfactant-nanoparticle hybrid system, the surfactant monomers direct themselves toward the surface of the nanoparticles through interaction of the hydrophobic surfactant tail with the nanoparticles *via* electrostatic interaction or van der Waals forces during the synthesis process, while the hydrophilic surfactant head is in contact with the polar solvent, consequently reducing the interfacial tension between the nanoparticles and solvent. This arrangement decreases the free energy of the surfactant-nanoparticle hybrid system and enhances stabilization of the nanoparticles in the surfactant solution.[83–87]

The stability is also influenced by the hydrophobic tail of the surfactant and the spacer, as in the case of the Gemini surfactant, where these factors improve the stability of the produced nanofluid system. In general, modification of the chain length of surfactants and the hydrophobic spacers between the hydrophilic heads are found to control the size distribution of the as-synthesized nanoparticles, which can provide a route for enhancing

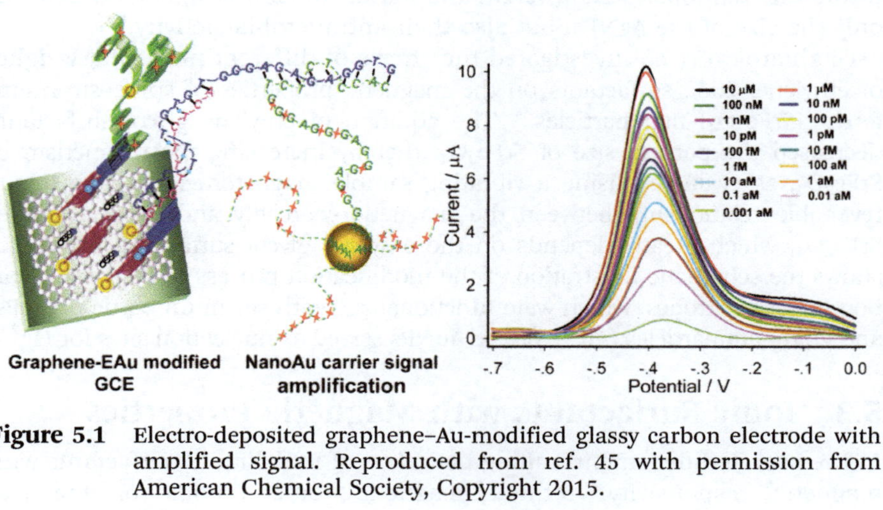

Figure 5.1 Electro-deposited graphene–Au-modified glassy carbon electrode with amplified signal. Reproduced from ref. 45 with permission from American Chemical Society, Copyright 2015.

the sensitivity of nanoscale sensors. Adsorption of surfactant molecules onto the surface of the nanomaterials not only controls their stability, but also significantly controls the particle size. The surfactant is characterized by excellent adsorption on the surface of the nanoparticles and can readily form micelles. The presence of the surfactant during nucleation of the nanoparticles can control the growth of the nanoparticles. As the hydrophobicity of the surfactant increases, the rate of migration onto the surfaces of nanoparticles in nanomaterials increases, forming a more protective layer, thereby maintaining the dispersion of the nanoparticles in solution with less agglomeration. Consequently, a larger particle size is attained.

5.2.2 Surfactant Structure-dependent Nanoparticle Properties

Modification of the geometry of as-synthesized nanoparticles based on the molecular structures of the surfactants, where the size distribution and the colloidal stability are strongly correlated to the alkyl chain length, can open a new avenue for optimizing the desired properties of nanoparticles. In this section, we discuss the effects of varying the surfactant structure (*i.e.*, alkyl chain length) on the nanoparticle properties. Dadwal and Joy *et al.* reported the colloidal stability of Fe_3O_4 nanoparticles in toluene solvent using myristic, palmitic, and stearic acid as surfactants with long alkyl chains. It was demonstrated that different chain lengths influenced the thermal conductivity of the magnetite nanofluid under an applied magnetic field.[88] The thermal conductivity decreased as the length of the surfactant chain increased.

Modification of the alkyl chain length also affects the antimicrobial activity of nanoparticles, as reported by Shaban *et al.* It was demonstrated that increasing the tail length of the surfactant changed the antibacterial activity of the AgNP nanohybrid system. In another study by Aiad *et al.*, a similar result was demonstrated wherein the surfactant tail length modified not only the size of the AgNPs, but also their antimicrobial activity.[89]

Ghahfarokhi *et al.* investigated the effects of different molecular weights of ethylene oxide surfactants on the magnetic properties of spinel-strontium ferrite ($SrFe_2O_4$) nanoparticles.[90] The addition of ethylene glycol surfactants decreased the particle size of $SrFe_2O_4$, thereby increasing the magnetism of $SrFe_2O_4$, as evaluated using a vibrating sample magnetometer, owing to the reversible relationship between the magnetic coercivity and particle size of $SrFe_2O_4$, which greatly depends on the ethylene glycol surfactant. Figure 5.2 shows the schematic illustration of the modification process of disposable carbon paper electrodes, which were functionalized with selenium particles (SePs) and gold nanoparticles (AuNPs). The AuNPs served as nucleation sites for Hg^{2+}.

5.3 Ionic Surfactants with Magnetic Properties

In the past decade, an enthralling new class of ionic liquid surfactants with magnetic responsivity, termed magnetic ionic liquids or magneto

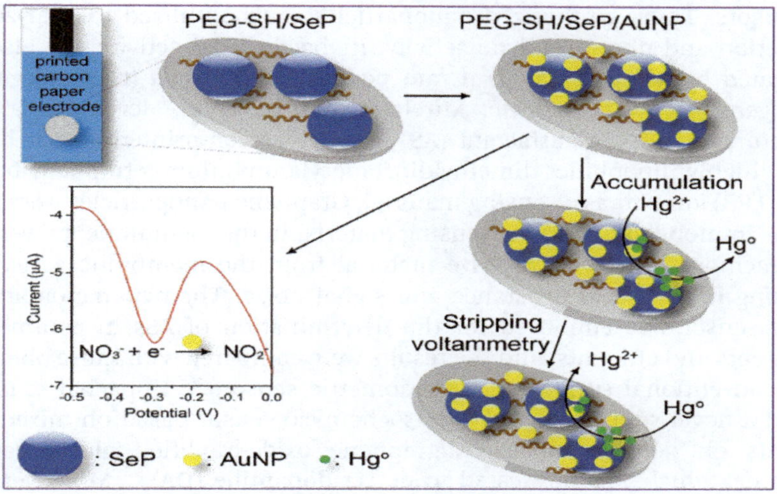

Figure 5.2 Schematic illustration of the modification process of disposable carbon paper electrodes, which were functionalized with selenium particles (SePs) and gold nanoparticles (AuNPs). The AuNPs served as nucleation sites for Hg^{2+}, whilst the SePs further reinforced the mercury ion nucleation due to their high binding affinity to mercury. Reproduced from ref. 50 with permission from Elsevier, Copyright 2016.

surfactants, has been reported. These new categories of surfactants exhibit a unique magnetic response in addition to the common properties of the surfactants (*i.e.*, adsorption and micellization).

The synthesized magneto surfactants were functionalized with proteins; consequently, protein separation and extraction was enabled by simply using an external magnetic field. It is noted that the toxicity of magneto surfactants, in some cases, is lower than that of conventional halide-based surfactant analogs which may make magneto surfactants suitable for use in contaminated or infected sites to enable selective microbial reduction without influencing normal tissues and cells.

5.3.1 Applications of Magneto Surfactants

For cancer therapy, the emerging technology of smart nano-vehicles has received significant attention for improving chemotherapy efficiency and preventing side effects. This has been increased by directing drugs to the targeted area and controlling drug release. Nucleic acids are crucial biological building blocks as they are characterized by molecular self-recognition.

Rajaboopathi *et al.* investigated a greener synthesis of Ag nanoparticles intercalated with ZnO particles, with seaweed as the biosurfactant component and silver nitrate, zinc nitrate, and sodium hydroxide as the precursors. Thus, synthesized nanoparticles can be utilized to analyze UV protection and photo catalytic activity. Its bactericidal activity was also investigated by using *S. aureus* (Gram positive) and *E. coli* (Gram negative) pathogens.[91] Olivera Galović, Mirela Samardžić *et al.* developed a new miniaturized anionic surfactant (AS)-selective screen-printed sensor based on a highly lipophilic dimethyldioctadecylammonium-tetraphenylborate (DDA-TPB) ion-pair as a sensing material. Graphene nanoparticles were used for the immobilization of the sensing material in the membrane, preventing the leaching of the electroactive material from the membrane as well as reducing its electrical resistance and signal noise. The new nanomaterial-based sensor was employed for the determination of ASs in commercial detergents and effluents, and the results were compared with those obtained with conventional surfactant potentiometric sensors.[92] Uppachai *et al.* revealed a novel supramolecular electrochemical sensor based on mixed surfactants on gold nanoparticles/graphene oxide-modified glassy carbon electrode, which was fabricated to detect dopamine (DA).[93] Moreover, the supramolecular electrochemical sensor showed a high level of sensitivity and excellent performance in stability and reproducibility. The proposed supramolecular electrochemical sensor was applied to human serum for the determination of DA. Shaban developed a colorimetric detection platform for citrate using AgNPs. The AgNPs are capped with two surfactants: cetyl trimethyl ammonium bromide (CTAB) having a positive quaternary nitrogen and a newly synthesized gemini (GFEO) surfactant containing multiple carbonyls and polyethylene oxide. This dual-surfactant system enabled highly sensitive and selective colorimetric detection of citrate. The limit of detection (LOD) of citrate was 4.05 nM in deionized (DI) water. The limit of visual detection by means of naked eye observation was 25 μM. Furthermore, the detection of citrate in urine samples was tested for practical applications, showing an LOD of 14.4 nM.[94]

5.3.2 Saccharide-based Biosensors

It is known that type 1 and type 2 diabetes is linked with saccharide content (*e.g.*, glucose) in the human system, and causes diseases such as heart failure, blindness, kidney damage, tissue degeneration, and even death. 3D-structured graphene has been applied to detect various saccharides such as sucrose, fructose, galactose, maltose, glucose, and others. Surfactant-based functional hybrid materials have been applied to detect various saccharides such as glucose, fructose, lactose, galactose, xylose, maltose, sucrose, and others. Bacteria, fungi, and viruses are causes of serious diseases.[95] It is essential to ascertain the blood's glucose level. The aforementioned virus can be supported by the appearance of graphene in sensitive and affordable biosensors.[96] Wang *et al.*[97] used graphene sheets as electrode material for developing glucose sensors. Various graphene-functionalized metal and metal oxide nanoparticles such as Ni, Cu, Ag, Au, CuO, SiO_2, Co_3O_4, and $NiCO_2O_4$ have been well studied for analytical and sensing performance for

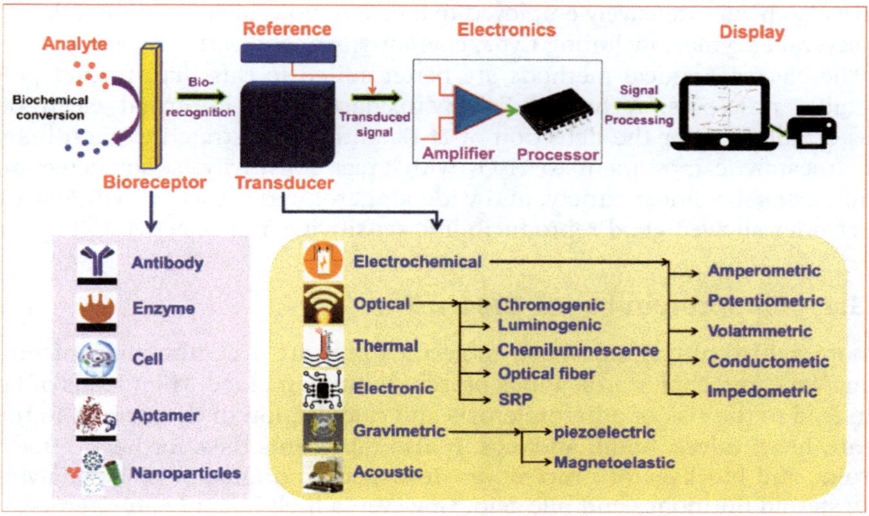

Figure 5.3 Schematic diagram of a typical biosensor consisting of bioreceptor, transducer, electronic system (amplifier and processor), and display (PC or printer) and various types of bioreceptors and transducers used in biosensors. Reproduced from ref. 61, https://doi.org/10.3390/s21041109, under the terms of the CC BY 4.0 license https://creativecommons.org/licenses/by/4.0/.

glucose. The developed graphene electrode showed good sensitivity and reproducibility for detection of glucose.[98] Figure 5.3 shows the components of biosensors.

5.3.3 Cytochrome Biosensor

Cytochrome-*c* (Cyt-*c*) is a small metalloprotein that is present in the compartment between the inner and outer mitochondrial membranes, and functions to transport electrons between complexes III and IV of the respiratory system. Therefore, Cyt-*c* plays a key role in determining the cell apoptosis process. The electron transfer takes place between electrodes and Cyt-*c* and is getting increasing investigation consideration due to its biological functions. However, pure glassy carbon electrodes cannot detect Cyt-*c*.[99] Therefore, carbon-based surfactant materials have been used to modify their surface. It was reported that a novel graphene material consisting of graphene/Ru(II) metal complex–CeO NPs–chitosan can be an outstanding sensing platform for the construction of electrochemiluminescence (ECL) sensors.[100] The surface of the electrode was incubated with bovine serum albumin to block nonspecific binding sites. This apta-detector showed a wide dynamic range from 0.2 to 100 nM with a low sensing limit of 25 pM that is applicable to the analysis of Cyt-*c* in order to find early-stage apoptotic cells. The interactions between graphene and glucose oxidase (GO)-based materials have also been extensively employed to create nano vehicles and modulators of several enzymes, including Cyt-*c*, chymotrypsin, and serine protease.

The electrochemical methods are better suited to this than further predictable processes. A horseradish peroxidase (HRP)-improved electrode was established for the detection of H_2O_2 and demonstrated extraordinary electrocatalytic movement to H_2O_2 with great sensitivity, squat detection limit, extensive linear variety, and wide amperometric reaction. HRP-based electrodes showed good reproducibility, sensitivity, and selectivity.[98]

5.3.4 Lipoprotein-based Biosensors

Lipoprotein (Lp) contains large molecules that are a combination of proteins, fats, and cholesterol, and is present in human blood. High levels of Lp can lead to the risk of atherosclerosis and calcification in the arteries to the heart, heart valves, brain, and legs. It also represents risks for heart attack, stroke, and blockage of heart valves. It is also an originator of bioanalytes like steroid hormones and bile acid. However, a high level of cholesterol can cause heart disease, arthrosclerosis, and other diseases, so that it is very important to develop sensors for the detection of cholesterol. A sensitive amperometric biosensor was developed by coating Pt nanoparticles on the surface of a graphene electrode. The sensitivity and detection limit of the electrode for detection of cholesterol were 2.07 $\mu M\,cm^{-2}$, and 0.2 μM, respectively. The excellent sensitivity was achieved due to the synergistic effect of graphene with Pt nanoparticles.[99]

5.3.5 Iron-based Biosensors

Hemoglobin (Hgb) is a protein biomolecule that is a component of red blood cells and plays a key role in carrying oxygen from the respiratory organs (lungs) to other parts of the body. The structure of Hgb consists of four peptide chains and each chain is connected with a heme group that consists of an iron (Fe) chelated to a protoporphyrin ring. The determination of Hgb level in the blood is useful for the diagnosis of several diseases such as heart disease, anemia, and leukemia. A chitosan@graphene hybrid electrode has been developed and used as a sensor for the detection of Hgb. The response to Hgb using a glassy carbon electrode was in the range of 30–150 $mV\,s^{-1}$.[100] The influence of magnetic nanoparticles on Hgb exposure was studied. A H_2O_2 sensor with a lower detection limit has been developing using graphene-based electrodes.[101] Zhan *et al.*[102] reported the fabrication of a 3D graphene/carbon nitride/layered double hydroxide hybrid (Gr-$LDHC_3N_4$) by a hydrothermal-calcination process. This hybrid and chitosan were used to immobilize Hgb on a carbon ionic liquid electrode for the fabrication of a trichloroacetic acid biosensor. It was reported that this electrode showed a wide linear sensing range from 0.2 to 36 mM with a low detection limit (DL) of 0.05 mM. Due to the effective electron transfer ability, large surface area of the conductive Gr–C_3N_4 hybrid, and the good biocompatibility of chitosan, the constructed sensor showed short response time, good stability, and high sensitivity.

5.3.6 Electrochemical Sensors for the Detection of Various Chemicals

Electrochemical sensing is an effective technique for the quick detection of various chemicals and biochemical species in various samples such as water, air, and food, as these sensors offer low cost, easy operation, and excellent sensitivity. Figure 5.4 shows a typical electrochemical measurement setup with three electrodes. The surfactant-based electrode materials show excellent characteristics such as controllable size, morphology, electrical conductivity, mechanical properties, and template suitability for the dispersion of dopants and modifying agents. Surfactant-based oxide hybrid electrodes were studied for sensing various chemicals such as alcohols, chloroform, phenols, acetone, different volatile organic compounds, and gases [*e.g.*, NO_2, H_2S, NH_3, CO, and LPG (liquified petroleum gas)]. A one-step hydrothermal method was employed to prepare GO coated with ZnO nanoclusters, used as an electrochemical sensor for the detection of bisphenol A.[103] The electrode had a detection limit of 2.1 nM. Such hybrid electrodes can also be applied for the detection of bisphenols present in food products such as milk,

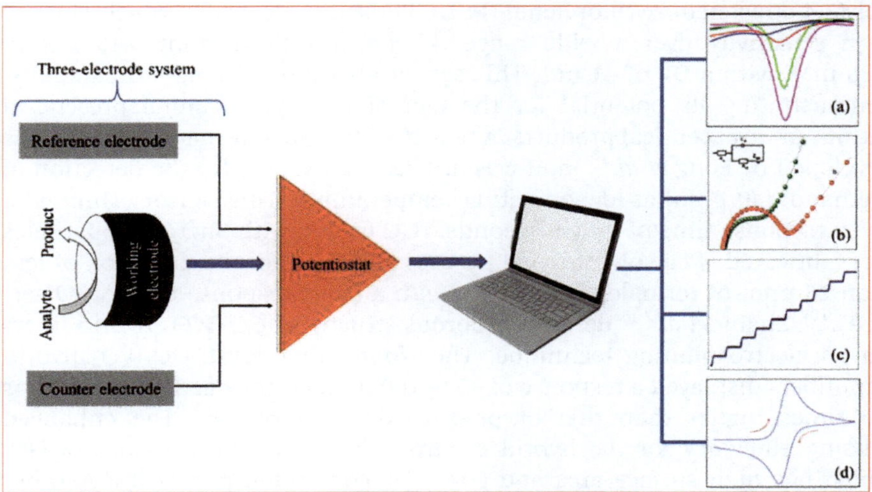

Figure 5.4 Typical electrochemical measurement setup with three electrodes, a working electrode, a reference electrode, and a counter electrode for (a) square wave anodic strip voltammetry (SWASV), (b) electrochemical impedance spectroscopy (EIS), (c) amperometric, and (d) cyclic voltammetry (CV). Reproduced from ref. 75, http://dx.doi.org/10.5599/jese.1506, under the terms of the CC BY 4.0 license, https://creativecommons.org/licenses/by/4.0/.

Table 5.1 Comparison of the proposed sensor with the reported methods for the detection of cadmium and mercuric ions.

Metal ions	Electrode substrate	Measurement technique	LOD ($mg L^{-1}$)	Ref.
Hg^{2+}	CDTGCE	SWASV	0.090	109
Cd^{2+}	CDTGCE	SWASV	0.084	109
Hg^{2+}	Biotinyl somatostatin-14/Nafion-modified gold electrode	Cyclic voltammetry	0.4	110
Cd^{2+}	G-MWCNTs	DPASV	0.1	111
Hg^{2+}	WBMCPE	DPASV	195	112
Hg^{2+}	SBA-15 with 2-benzothiazolethiol	DPASV	100	113
Cd^{2+}	CS-MCPE	DPASV	105	114
Hg^{2+}	SH-SAMMSCPE	SWASV	3	115
Hg^{2+}	DANGCE	SWASV	0.64	116

among others. Li *et al.*[104] prepared reduced graphene oxide (rGO) anionic metalorganic framework hybrids *via* a cationic exchange and reduction process, and investigated the hybrid electrode for the determination of H_2O_2 released from living cells. They observed that the electrocatalytic properties of the electrode depended on the content of the anionic metalorganic frameworks. The hybrid electrode showed a linear detection range from 4 to 11 000 μM with a DL of 0.18 μM. Li *et al.*[105] developed a sensitive electrochemical sensor for the detection of paracetamol using an Au–graphene@poly(3,4-ethylenedioxythiophene) (PEDOT) ternary electrode, which showed high sensitivity over a wide range of concentrations from 0.15 μM to 5.88 mM, with a DL of 41 nM. This sensor also showed long-term stability, demonstrating its potential for the detection of paracetamol present in various pharmaceutical products. A new rGO–Pt@SnO_2 ternary electrode was developed by Peng *et al.*[106] and was utilized as a sensor for the detection of methanol (40 ppm) at low operating temperatures. This sensor exhibited a short response time of only 6 seconds. rGO coated with SnO_2 nanoparticles was fabricated *via* a solvothermal process and used for the detection of less than 25 ppm of formaldehyde vapor, with a short response time of 20 seconds.[107] Zhang *et al.*[108] developed porous structured GO–WO_3 hybrid fibers *via* an electrospinning technique. They found that 1 mL GO-WO_3 hybrid nanofibers displayed a response of 35.9–100 ppm acetone at 375 °C, which is 4.3 times higher than that of pristine WO_3 nanofibers. The enhanced sensing efficiency for the hybrid electrode is due to the presence of GO, which has high surface area and good gas adsorption properties. Another new type of hybrid electrode was developed based on rGO–GaN nanorods through a spin-coating process and used for the sensing of H_2 and H_2S. It was found that the hybrid electrode showed a higher response to those gases compared with pure GaN. This hybrid electrode exhibited excellent sensing performance at room temperature under UV light. Various surfactant-based hybrid functional materials have been researched for the detection of various gases and chemicals. Table 5.1 shows the comparison of the proposed sensor with the reported methods for the detection of cadmium and mercuric ions.

5.4 Conclusion and Future Outlook

This chapter has presented surfactant-based electrode materials that were fabricated through different strategies and used as electrochemical sensors and biosensors. Different approaches have been discovered intended for the fabrication of different biosensors based on surfactant-based graphene or other nanocarbon resources as essentials for electrochemical sensing; however, there is still a need for further investigation. We have discussed the recent advances in surfactants and their tunable, structure-dependent properties for optimized applications in nanotechnology. During the synthesis of nanoparticles in solution, surfactants play an important role in improving the dispersion stability and controlling the size distributions of the as-synthesized nanoparticles. More specifically, as the tail length of the surfactant used for nanoparticle synthesis increases, smaller, less aggregated, and more stable nanoparticles are produced. This development can significantly promote electrochemical and biosensor applications based on size-tunable nanoparticles. However, the correlation between the surfactant structure and the nanoparticle geometry should still be defined through intensive studies to provide direct evidence of this correlation and molecular fingerprints. Furthermore, a new category of surfactants, namely, magneto surfactants, has been introduced. By forming a mixture of conventional surfactants and magnetic elements, a new series of surfactants having unique magnetic properties can be successfully synthesized. Magneto surfactants have been further utilized in the field of drug delivery because of their high selectivity and sensitivity to target a certain position. Although the emerging category of surfactants has been intensively studied, it is still necessary to elucidate the fundamental mechanisms, provide technical solutions, achieve deeper understanding of their toxicity, and enable the possibility of large-scale production for commercial uses, considering the short history of surfactants since their initial discovery.

References

1. N. Yekeen, *et al.*, Influence of surfactant and electrolyte concentrations on surfactant Adsorption and foaming characteristics, *J. Pet. Sci. Eng.*, 2017, **149**, 612–622.
2. D. Zhang, *et al.*, Synthesis and properties study of novel fluorinated surfactants with perfluorinated branched ether chain, *J. Fluorine Chem.*, 2019, **219**, 62–69.
3. M. S. Alam, *et al.*, Physicochemical properties and bioactivity studies of synthesized counterion coupled (COCO) gemini surfactant, 1,6-bis(N,N-hexadecyldimethylammonium) adipate, *J. Mol. Liq.*, 2019, **273**, 16–26.

4. S. M. S. Hussain, M. S. Kamal and L. T. Fogang, Synthesis and physicochemical investigation of betaine type polyoxyethylene zwitterionic surfactants containing different ionic headgroups, *J. Mol. Struct.*, 2019, **1178**, 83–88.
5. N. Pal, K. Samanta and A. Mandal, A novel family of non-ionic gemini surfactants derived from sunflower oil: Synthesis, characterization and physicochemical evaluation, *J. Mol. Liq.*, 2019, **275**, 638–653.
6. A. R. Tehrani-Bagha, Cationic gemini surfactant with cleavable spacer: Emulsion stability, *Colloids Surf., A*, 2016, **508**, 79–84.
7. S. M. Shaban, *et al.*, Some alginates polymeric cationic surfactants; surface study and their evaluation as biocide and corrosion inhibitors, *J. Mol. Liq.*, 2019, **273**, 164–176.
8. F. Zhang, *et al.*, Adsorption of different types of surfactants on graphene oxide, *J. Mol. Liq.*, 2019, **276**, 338–346.
9. M. G. Chernysheva, *et al.*, Cationic surfactant coating nanodiamonds: Adsorption and peculiarities, *Colloids Surf., A*, 2019, **565**, 25–29.
10. Z. Chang, X. Chen and Y. Peng, The adsorption behavior of surfactants on mineral surfaces in the presence of electrolytes – A critical review, *Miner. Eng.*, 2018, **121**, 66–76.
11. A. Casandra, *et al.*, Adsorption kinetics of the partially dissociated ionic surfactants: The effect of degree of dissociation, *J. Taiwan Inst. Chem. Eng.*, 2018, **92**, 2–7.
12. D. Möbius, R. Miller and V. B. Fainerman, *Surfactants: chemistry, interfacial properties, applications*, Elsevier, 2001, vol. 13.
13. M. J. Rosen, *Surfactants and Interfacial Phenomena*, Wiley, 2004.
14. D. R. Karsa, *Industrial applications of surfactants IV*, Elsevier, 1999.
15. J. G. Clarke, S. R. Wicks and S. J. Farr, Surfactant mediated effects in pressurized metered dose inhalers formulated as suspensions. I. Drug/surfactant interactions in a model propellant system, *Int. J. Pharm.*, 1993, **93**(1), 221–231.
16. V. P. Torchilin, Structure and design of polymeric surfactant-based drug delivery systems, *J. Controlled Release*, 2001, **73**(2), 137–172.
17. M. El Achouri, *et al.*, Corrosion inhibition of iron in 1 M HCl by some gemini surfactants in the series of alkanediyl-α,ω-bis-(dimethyl tetradecyl ammonium bromide), *Prog. Org. Coat.*, 2001, **43**(4), 267–273.
18. M. A. Hegazy, M. Abdallah and H. Ahmed, Novel cationic gemini surfactants as corrosion inhibitors for carbon steel pipelines, *Corros. Sci.*, 2010, **52**(9), 2897–2904.
19. M. Elachouri, *et al.*, Some surfactants in the series of 2-(alkyldimethylammonio) alkanol bromides as inhibitors of the corrosion of iron in acid chloride solution, *Corros. Sci.*, 1995, **37**(3), 381–389.
20. L.-G. Qiu, *et al.*, Synergistic effect between cationic gemini surfactant and chloride ion for the corrosion inhibition of steel in sulphuric acid, *Corros. Sci.*, 2008, **50**(2), 576–582.

21. H.-Y. Wu, *et al.*, Development and validation of an analytical procedure for quantitation of surfactants in dishwashing detergents using ultra-performance liquid chromatography-mass spectrometry, *Talanta*, 2019, **194**, 778–785.
22. S. Yada, *et al.*, Emulsification, solubilization, and detergency behaviors of homogeneous polyoxypropylene-polyoxyethylene alkyl ether type nonionic surfactants, *Colloids Surf., A*, 2019, **564**, 51–58.
23. S. Lee, *et al.*, Synthesis of environment friendly nonionic surfactants from sugar base and characterization of interfacial properties for detergent application, *J. Ind. Eng. Chem.*, 2016, **38**, 157–166.
24. A. M. Atta, *et al.*, Demulsification of heavy crude oil using new nonionic cardanol surfactants, *J. Mol. Liq.*, 2018, **252**, 311–320.
25. A. M. Abdulraheim, Green polymeric surface active agents for crude oil demulsification, *J. Mol. Liq.*, 2018, **271**, 329–341.
26. C. Xu, *et al.*, Experimental investigation of coal dust wetting ability of anionic surfactants with different structures, *Process Saf. Environ. Prot.*, 2019, **121**, 69–76.
27. L. Alexandrova, *et al.*, Effects of pH on wetting behavior of 'star-like' block copolymer surfactant solutions, *Colloids Surf., A*, 2017, **519**, 78–86.
28. Y. Wang, *et al.*, Halide-free PN ionic liquids surfactants as additives for enhancing tribological performance of water-based liquid, *Tribol. Int.*, 2018, **128**, 190–196.
29. M. Zhao, *et al.*, Study on the synergy between silica nanoparticles and surfactants for enhanced oil recovery during spontaneous imbibition, *J. Mol. Liq.*, 2018, **261**, 373–378.
30. A. M. Howe, *et al.*, Visualising surfactant enhanced oil recovery, *Colloids Surf., A*, 2015, **480**, 449–461.
31. E. Nourafkan, Z. Hu and D. Wen, Controlled delivery and release of surfactant for enhanced oil recovery by nanodroplets, *Fuel*, 2018, **218**, 396–405.
32. A. A. Hafiz and T. T. Khidr, Hexa-triethanolamine oleate esters as pour point depressant for waxy crude oils, *J. Pet. Sci. Eng.*, 2007, **56**(4), 296–302.
33. N. El Mehbad, Efficiency of N-Decyl-N-benzyl-N-methylglycine and N-Dodecyl-N-benzyl-Nmethylglycine surfactants for flow improvers and pour point depressants, *J. Mol. Liq.*, 2017, **229**, 609–613.
34. X. Gu, *et al.*, Investigation of cationic surfactants as clean flow improvers for crude oil and a mechanism study, *J. Pet. Sci. Eng.*, 2018, **164**, 87–90.
35. Z. Weiszhár, *et al.*, Complement activation by polyethoxylated pharmaceutical surfactants: Cremophor-EL, Tween-80 and Tween-20, *Eur. J. Pharm. Sci.*, 2012, **45**(4), 492–498.
36. D. Steinhilber, *et al.*, Surfactant free preparation of biodegradable dendritic polyglycerol nanogels by inverse nanoprecipitation for encapsulation and release of pharmaceutical biomacromolecules, *J. Controlled Release*, 2013, **169**(3), 289–295.

37. K. Bouchemal, *et al.*, Nano-emulsion formulation using spontaneous emulsification: solvent, oil and surfactant optimisation, *Int. J. Pharm.*, 2004, **280**(1), 241–251.
38. C. J. Drummond and C. Fong, Surfactant self-assembly objects as novel drug delivery vehicles, *Curr. Opin. Colloid Interface Sci.*, 1999, **4**(6), 449–456.
39. I. F. Uchegbu and S. P. Vyas, Non-ionic surfactant based vesicles (niosomes) in drug delivery, *Int. J. Pharm.*, 1998, **172**(1), 33–70.
40. A. Rivera and T. Farías, Clinoptilolite–surfactant composites as drug support: A new potential application, *Microporous Mesoporous Mater.*, 2005, **80**(1), 337–346.
41. Q. Zhao, *et al.*, Interaction of nano carbon particles and anthracene with pulmonary surfactant: The potential hazards of inhaled nanoparticles, *Chemosphere*, 2019, **215**, 746–752.
42. M. Hasheminejad and A. Nezamzadeh-Ejhieh, A novel citrate selective electrode based on surfactant modified nano-clinoptilolite, *Food Chem.*, 2015, **172**, 794–801.
43. B. Huang, *et al.*, Size-controlled synthesis and morphology evolution of Nd_2O_3 nano-powders using ionic liquid surfactant templates, *J. Alloys Compd.*, 2017, **712**, 164–171.
44. W. Amer, *et al.*, Synthesis of mesoporous nano-hydroxyapatite by using zwitterions surfactant, *Mater. Lett.*, 2013, **107**, 189–193.
45. Y. Zhang, G. M. Zeng, L. Tang, J. Chen, Y. Zhu, X. X. He and Y. He, Electrochemical sensor based on electrodeposited graphene-Au modified electrode and nanoAu carrier amplified signal strategy for attomolar mercury detection, *Anal. Chem.*, 2015, **87**, 989–996.
46. N. Asikin-Mijan, Y. H. Taufiq-Yap and H. V. Lee, Synthesis of clamshell derived $Ca(OH)_2$ nanoparticles via simple surfactant-hydration treatment, *Chem. Eng. J.*, 2015, **262**, 1043–1051.
47. M. C. Hersam, Progress towards monodisperse single-walled carbon nanotubes, *Nat. Nanotechnol.*, 2008, **3**(7), 387–394.
48. X. Wang, *et al.*, Toxicological Profiling of Highly Purified Single-Walled Carbon Nanotubes with Different Lengths in the Rodent Lung and Escherichia Coli, *Small*, 2018, **14**(23), 1703915.
49. K. Yoon, *et al.*, Metal-Free Carbon-Based Nanomaterial Coatings Protect Silicon Photoanodes in Solar Water-Splitting, *Nano Lett.*, 2016, **16**(12), 7370–7375.
50. M.-P. N. Bui, J. Brockgreitens, S. Ahmed and A. Abbas, Dual detection of nitrate and mercury in water using disposable electrochemical sensors, *Biosens. Bioelectron.*, 2016, **85**, 280–286.
51. J. Kang, *et al.*, Thickness sorting of two-dimensional transition metal dichalcogenides via copolymer-assisted density gradient ultracentrifugation, *Nat. Commun.*, 2014, **5**(1), 5478.
52. J. Kang, *et al.*, Solvent Exfoliation of Electronic-Grade, Two-Dimensional Black Phosphorus, *ACS Nano*, 2015, **9**(4), 3596–3604.
53. K. S. Novoselov, *et al.*, Electric Field Effect in Atomically Thin Carbon Films, *Science*, 2004, **306**(5696), 666.

54. J. Kang, *et al.*, Solution-Based Processing of Monodisperse Two-Dimensional Nanomaterials, *Acc. Chem. Res.*, 2017, **50**(4), 943–951.
55. G. Hu, *et al.*, Functional inks and printing of two-dimensional materials, *Chem. Soc. Rev.*, 2018, **47**(9), 3265–3300.
56. J. Kang, *et al.*, Solution-Processed Layered Gallium Telluride Thin-Film Photodetectors, *ACS Photonics*, 2018, **5**(10), 3996–4002.
57. J. Kang, *et al.*, Solution-Based Processing of Optoelectronically Active Indium Selenide, *Adv. Mater.*, 2018, **30**(38), 1802990.
58. D. Lam, *et al.*, Anhydrous Liquid-Phase Exfoliation of Pristine Electrochemically Active GeS Nanosheets, *Chem. Mater.*, 2018, **30**(7), 2245–2250.
59. J. Kang, *et al.*, Stable aqueous dispersions of optically and electronically active phosphorene, *Proc. Natl. Acad. Sci. U. S. A.*, 2016, **113**(42), 11688–11693.
60. S. Kim, *et al.*, Carbon nanotube ferroelectric random access memory cell based on omega shaped ferroelectric gate, *Carbon*, 2020, **162**, 195–200.
61. V. Naresh and N. Lee, A review on biosensors and recent development of nanostructured materials-enabled biosensors, *Sensors*, 2021, **21**, 1109.
62. Y. Choi, *et al.*, Capacitively Coupled Hybrid Ion Gel and Carbon Nanotube Thin-Film Transistors for Low Voltage Flexible Logic Circuits, *Adv. Funct. Mater.*, 2018, **28**(34), 1802610.
63. C. Zhong, *et al.*, Hot Carrier and Surface Recombination Dynamics in Layered InSe Crystals, *J. Phys. Chem. Lett.*, 2019, **10**(3), 493–499.
64. C. Husko, *et al.*, Silicon-Phosphorene Nanocavity-Enhanced Optical Emission at Telecommunications Wavelengths, *Nano Lett.*, 2018, **18**(10), 6515–6520.
65. M. Engel, *et al.*, Graphene-enabled and directed nanomaterial placement from solution for large-scale device integration, *Nat. Commun.*, 2018, **9**(1), 4095.
66. Q. Li, *et al.*, Revealing the Effects of Electrode Crystallographic Orientation on Battery Electrochemistry via the Anisotropic Lithiation and Sodiation of ReS_2, *ACS Nano*, 2018, **12**(8), 7875–7882.
67. L. H. Guo, *et al.*, Strategies for enhancing the sensitivity of plasmonic nanosensors, *Nano Today*, 2015, **10**(2), 213–239.
68. P. K. Jain, *et al.*, Noble metals on the nanoscale: optical and photothermal properties and some applications in imaging, sensing, biology, and medicine, *Acc. Chem. Res.*, 2008, **41**(12), 1578–1586.
69. M. E. Stewart, *et al.*, Nanostructured Plasmonic Sensors, *Chem. Rev.*, 2008, **108**(2), 494–521.
70. S. Song, *et al.*, Functional nanoprobes for ultrasensitive detection of biomolecules, *Chem. Soc. Rev.*, 2010, **39**(11), 4234–4243.
71. P. D. Howes, R. Chandrawati and M. M. Stevens, Colloidal nanoparticles as advanced biological sensors, *Science*, 2014, **346**(6205), 1247390.
72. S. Choi, *et al.*, Recent Advances in Flexible and Stretchable Bio-Electronic Devices Integrated with Nanomaterials, *Adv. Mater.*, 2016, **28**(22), 4203–4218.

73. F. Wu, P. Yu and L. Mao, Self-powered electrochemical systems as neurochemical sensors: toward self-triggered in vivo analysis of brain chemistry, *Chem. Soc. Rev.*, 2017, **46**(10), 2692–2704.
74. P. Kumar, *et al.*, Hybrid porous thin films: Opportunities and challenges for sensing applications, *Biosens. Bioelectron.*, 2018, **104**, 120–137.
75. S. N. A. M. Yazid, A. A. C. Adnan, I. M. Isa, M. I. Saidin, M. S. Ahmad and C. S. Fun, Conducting polymer functionalized graphene-based electrochemical sensors for sensing pollutants in water, *J. Electrochem. Sci. Eng.*, 2023, **13**(2), 251–274.
76. T. Gunnlaugsson, *et al.*, Colorimetric "Naked Eye" Sensing of Anions in Aqueous Solution, *J. Org. Chem.*, 2005, **70**(26), 10875–10878.
77. Y. Song, W. Wei and X. Qu, Colorimetric Biosensing Using Smart Materials, *Adv. Mater.*, 2011, **23**(37), 4215–4236.
78. M. Rana and P. Chowdhury, L-glutathione capped CdSeS/ZnS quantum dot sensor for the detection of environmentally hazardous metal ions, *J. Lumin.*, 2019, **206**, 105–112.
79. N. Xiao, *et al.*, Multifunctional fluorescent sensors for independent detection of multiple metal ions based on Ag nanoclusters, *Sens. Actuators, B*, 2018, **264**, 184–192.
80. W. Jiang, *et al.*, A novel fluorescence "turn off-on" nano-sensor for detecting Cu^{2+} and Cysteine in living cells, *J. Photochem. Photobiol., A*, 2018, **362**, 14–20.
81. R. Salahandish, *et al.*, Nano-biosensor for highly sensitive detection of HER2 positive breast cancer, *Biosens. Bioelectron.*, 2018, **117**, 104–111.
82. H. Yang, *et al.*, A cancer cell turn-on protein-CuSMn nanoparticle as the sensor of breast cancer cell and CH_3O-PEG-phosphatide, *Chin. Chem. Lett.*, 2018, **29**(10), 1528–1532.
83. O. Naderi, *et al.*, Synthesis and characterization of silver nanoparticles in aqueous solutions of surface active imidazolium-based ionic liquids and traditional surfactants SDS and DTAB, *J. Mol. Liq.*, 2019, **273**, 645–652.
84. J. Yu, *et al.*, Controlling the dispersion of multi-wall carbon nanotubes in aqueous surfactant solution, *Carbon*, 2007, **45**(3), 618–623.
85. H. Wang, *et al.*, Dispersing Single-Walled Carbon Nanotubes with Surfactants: A Small Angle Neutron Scattering Study, *Nano Lett.*, 2004, **4**(9), 1789–1793.
86. S. M. Shaban, *et al.*, Studying surface and thermodynamic behavior of a new multi-hydroxyl Gemini cationic surfactant and investigating their performance as corrosion inhibitor and biocide, *J. Mol. Liq.*, 2020, **316**, 113881.
87. S. M. Shaban and D.-H. Kim, The influence of the Gemini surfactants hydrocarbon tail on in-situ synthesis of silver nanoparticles: Characterization, surface studies and biological performance, *Korean J. Chem. Eng.*, 2020, **37**(6), 1008–1019.

88. A. Dadwal and P. A. Joy, Influence of chain length of long-chain fatty acid surfactant on the thermal conductivity of magnetite nanofluids in a magnetic field, *Colloids Surf., A*, 2018, **555**, 525–531.
89. I. Aiad, *et al.*, Antipyrine cationic surfactants capping silver nanoparticles as potent antimicrobial agents against pathogenic bacteria and fungi, *J. Mol. Liq.*, 2017, **243**, 572–583.
90. S. E. Mousavi Ghahfarokhi and E. Mohammadzadeh Shobegar, An investigation of the ethylene glycol surfactant on the structural, microstructure, magnetic and optical properties of $SrFe_2O_4$ nanoparticles, *J. Magn. Magn. Mater.*, 2020, **495**, 165866.
91. S. Rajaboopathi and S. Thambidurai, Synthesis of bio-surfactant based Ag/ZnO nanoparticles for better thermal, photocatalytic and antibacterial activity, *Mater. Chem. Phys.*, 2019, **223**, 512–522.
92. O. Galović, M. Samardžić, M. Hajduković and M. Sak-Bosnar, A New Graphene-based Surfactant Sensor for the Determination of Anionic Surfactants in Real Samples, *Sens. Actuators, B*, 2016, **236**, 257–267.
93. P. Uppachai, S. Srijaranai, S. Poosittisak, I. Md Isa and S. Mukdasai, Supramolecular electrochemical sensor for dopamine detection based on self-assembled mixed surfactants on gold nanoparticles deposited graphene oxide, *Molecules*, 2020, **25**(11), 2528.
94. (a) S. M. Shaban, J. Y. Lee and D. H. Kim, Dual-surfactant-capped Ag nanoparticles as a highly selective and sensitive colorimetric sensor for citrate detection, *ACS Omega*, 2020, **5**(19), 10696–10703; (b) M. S. Artiles, C. S. Rout and T. S. Fisher, Graphene-based hybrid materials and devices for biosensing, *Adv. Drug Delivery Rev.*, 2011, **63**, 1352–1360.
95. Y. Wang, Nitrogen-doped graphene and its application in electrochemical biosensing, *ACS Nano*, 2010, **4**, 1790–1798.
96. Z. Wang, X. Zhou, J. Zhang, F. Boey and H. Zhang, Direct electrochemical reduction of single-layer graphene oxide and subsequent functionalization with glucose oxidase, *J. Phys. Chem. C*, 2009, **113**, 14071–14075.
97. M. H. Yang, B. G. Choi, H. Park, W. H. Hong, S. Y. Lee and T. J. Park, Development of a glucose biosensor using advanced electrode modified by nanohybrid composing chemically modified graphene and ionic liquid, *Electroanalysis*, 2010, **22**, 1223–1228.
98. C. Xiang, Y. Zou, L.-X. Sun and F. Xu, Direct electron transfer of cytochrome *c* and its biosensor based on gold nanoparticles/room temperature ionic liquid/carbon nanotubes composite film, *Electrochem. Commun.*, 2008, **10**, 38–41.
99. J. F. Wu, M. Q. Xu and G. C. Zhao, Graphene-based modified electrode for the direct electron transfer of Cytochrome *c* and biosensing, *Electrochem. Commun.*, 2010, **12**, 175–177.
100. M. R. Pur, M. Hosseini, F. Faridbod, M. R. Ganjali and S. Hosseinkhani, *Sens. Actuators, B*, 2018, **257**, 87–95.

101. J. Wang and Y. Lin, Functionalized carbon nanotubes and nanofibers for biosensing applications, *TrAC, Trends Anal. Chem.*, 2008, **27**, 619–626.
102. T. Zhan, Z. Tan, X. Wang and W. Hou, Hemoglobin immobilized in g-C_3N_4 nanoparticle decorated 3D graphene-LDH network: direct electrochemistry and electrocatalysis to trichloroacetic acid, *Sens. Actuators, B*, 2018, **255**, 149–158.
103. C. Li, R. Wu, J. Zou, T. Zhang, S. Zhang and Z. Zhang, *et al.*, MNPs@anionic MOFs/ERGO with the size selectivity for the electrochemical determination of H_2O_2 released from living cells, *Biosens. Bioelectron.*, 2018, **116**, 81–88.
104. M. Li, W. Wang, Z. Chen, Z. Song and X. Luo, Electrochemical determination of paracetamol based on Au@graphene core-shell nanoparticles doped conducting polymer PEDOT nanocomposite, *Sens. Actuators, B*, 2018, **260**, 778.
105. R. Peng, J. Chen, X. Nie, D. Li, P. Si and J. Feng, *et al.*, Reduced graphene oxide decorated Pt activated SnO_2 nanoparticles for enhancing methanol sensing performance, *J. Alloys Compd.*, 2018, **762**, 8–15.
106. X. Rong, D. Chen, G. Qu, T. Li, R. Zhang and J. Sun, Effects of graphene on the microstructures of SnO_2@rGO nanocomposites and their formaldehyde sensing performance, *Sens. Actuators, B*, 2018, **269**, 223–237.
107. J. Zhang, H. Lu, C. Yan, Z. Yang, G. Zhu and J. Gao, *et al.*, Fabrication of conductive graphene oxide-WO_3 composite nanofibers by electrospinning and their enhanced acetone gas sensing properties, *Sens. Actuators, B*, 2018, **264**, 128–138.
108. M. Reddeppa, B. G. Park, M. D. Kim, K. R. Peta, N. D. Chinh and D. Kim, *et al.*, H_2, H_2S gas sensing properties of rGO/GaN nanorods at room temperature: effect of UV illumination, *Sens. Actuators, B*, 2018, **264**, 353–362.
109. A. Zahida, A. Shaha, F. J. Iftikhara, A. H. Shahb and R. Qureshia, Surfactant modified glassy carbon electrode as an efficient sensing platform for the detection of Cd (II) and Hg (II), *Electrochim. Acta*, 2017, **235**, 72–78.
110. N. Daud, N. A. Yusof and S. M. M. Nor, Electrochemical characteristic of biotinyl somatostatin-14/Nafion modified gold electrode in development of sensor for determination of Hg (II), *Int. J. Electrochem. Sci.*, 2013, **8**, 10086–10099.
111. H. Huang, T. Chen, X. Liu and H. Ma, Ultrasensitive and simultaneous detection of heavy metal ions based on three-dimensional graphene-carbon nanotubes hybrid electrode materials, *Anal. Chim. Acta*, 2014, **852**, 45–54.
112. D. S. Rajawat, S. Srivastava and S. P. Satsangee, Electrochemical determination of mercury at trace levels using eichhornia crassipes modified carbon paste electrode, *Int. J. Electrochem. Sci.*, 2012, **11**, 11456–11469.

113. I. Cesarino, G. M. R. Matos, J. Cavalheiro and E. T. G. Cavalheiro, Evaluation of a carbon paste electrode modified with organofunctionalised SBA-15 nanostructured silica in the simultaneous determination of divalent lead, copper and mercury ions, *Talanta*, 2008, **75**, 15–23.
114. D. S. Rajawat, N. Kumar and S. P. Satsangee, Trace determination of cadmium in water using anodic stripping voltammetry at a carbon paste electrode modified with coconut shell powder, *J. Anal. Sci. Technol.*, 2014, **5**, 19.
115. W. Yantasee, Y. Lin, T. S. Zemanian and G. E. Fryxell, Voltammetric detection of lead(II) and mercury(II) using a carbon paste electrode modified with thiol selfassembled monolayer on mesoporous silica (SAMMS), *Analyst*, 2003, **128**, 467–472.
116. A. Zahid, A. Lashin, U. A. Rana, N. Al-Arifi, I. Ullah, D. D. Dionysiou, R. Qureshi, A. Waseem, H. B. Kraatz and A. Shah, Development of surfactant based electrochemical sensor for the trace level detection of mercury, *Electrochim. Acta*, 2016, **190**, 1007–1014.

CHAPTER 6

Surfactant-based Potentiometric Sensors for Surfactant Determination

BRUNA COLDIBELI,[a] MARCO AURELIO JEANEGITZ CLEMENTE[b] AND ELEN ROMÃO SARTORI*[a]

[a] Laboratório de Eletroanalítica e Sensores, Departamento de Química, Centro de Ciências Exatas, Universidade Estadual de Londrina, Rodovia Celso Garcia Cid, PR 445 Km 380, CEP 86057-970, Londrina – PR, Brazil; [b] Laboratório de Pesquisa e Análises de Combustíveis, Departamento de Química, Centro de Ciências Exatas, Universidade Estadual de Londrina, CEP 86057-970, Londrina – PR, Brazil
*Email: elensartori@uel.br

6.1 Introduction

Various types of pollutants such as pharmaceuticals, pesticides, dyes, metals, and surfactants from cities and industry contribute to the pollution of the environment, especially water resources. Among these, surfactants are emerging contaminants used daily in large quantities by industry and households. The release of these persistent and water-soluble compounds into water bodies due to inadequate wastewater treatment can cause serious problems for the ecosystem. They can be responsible for changes in the physical and chemical parameters of the water, such as salinity, temperature, turbidity, and pH.[1,2] The consequences are damage to fish gills, an increase in the difficulty of breathing and swimming of the fishes, and a decrease in oxygen and light transmission in the water. In humans, when

Detection Science Series No. 23
Surfactant-based Sensors in Chemical and Biochemical Detection
Edited by J. G. Manjunatha

Published by the Royal Society of Chemistry, www.rsc.org

present in drinking or irrigation water, for example, surfactants can be responsible for various effects, such as disruption of endocrine function and the immune system.[3–5]

Surfactants are used, among other things, as wetting agents, detergents, and emulsifiers in various industrial segments such as textiles, pharmaceuticals, cosmetics, mining, oil recovery, and paper.[6] In 2021, more than three million tons of surfactants were produced in Western Europe, and total surfactant industry turnover was estimated at €4.4 billion. With the arrival of the COVID pandemic, the use of chemical products increased, and according to Alygizakis *et al.*,[7] who analyzed samples from wastewater treatment plants under lockdown conditions, surfactants increased by +196%.[8,9]

Surfactants have a high molecular weight and a complex structure, and their main function is to improve the cleaning power of liquids. This is done by reducing the surface tension between two liquids or between a liquid and a solid, which is made possible by the hydrophilic and lipophilic groups contained in these amphiphilic molecules.[6,8] The non-polar part of the molecule consists of a hydrocarbon chain, and the polar part can be divided into ionic (cationic and anionic) or non-ionic. Some examples of cationic surfactants are hexadecyltrimethylammonium bromide (CTAB), cetylpyridinium chloride (CPC), benzethonium chloride (BTC), *N*-(1,1,2,2-tetrahydroperfluorooctyl)-*N*,*N*-dimethylammonium chloride (TFACl), and of anionic surfactants are sodium dodecyl sulfate (SDS), sodium dodecylbenzene sulfonate, sodium tridecylsulfate, sodium tetradecylsulfate, and sodium hexadecylsulfate.[10]

In view of this, the determination of surfactants in both environmental and commercial samples is of great importance, and two well-known standard methods are used. The Methylene Blue Active Substance (MBAS) method is based on spectrophotometric measurements for the determination of surfactants at low concentrations, for example in environmental water samples. Its operation consists in the formation of an ion pair between the anionic surfactant from the sample and a cationic dye (methylene blue), followed by its extraction from the aqueous phase in chloroform. The ion pair formed is determined by spectrophotometry at 650 nm. Despite its simplicity and low cost, the MBAS method is a time-consuming process and requires significant amounts of sample and chloroform, which is a toxic solvent.[11,12]

On the other hand, high concentrations of anionic surfactants are determined by the traditional colorimetric method of two-phase titration. This determination can be performed in a water/chloroform system, where the end-point is found by the color change from green to blue of the layer with the organic solvent (chloroform) due to the complexation of the cationic titrant with the anionic surfactant. Solvatochromic dyes are used as end-point indicators.[13] Because an experienced analyst is required and toxic solvents are used, this technique is considered difficult to perform.[14,15]

In an effort to overcome the drawbacks of standard methods, analytical alternatives for surfactant monitoring have been extensively investigated, particularly in terms of cost effectiveness, speed, simplicity, and green chemistry. Thus, ion-selective electrodes (ISEs) are an excellent option for this purpose as they meet current analytical requirements. The development of novel ISEs for potentiometric titration of surfactants has shown several advantages in practice, offering high sensitivity and reliability, low toxic solvent consumption, no problems with colored and turbid samples, no need for sample preparation, and portable instrumentation of low-cost. In addition, their accuracy is comparable to that of standard methods.[16,17]

6.2 Surfactant-based Potentiometric Sensors for Determination of Surfactants

Potentiometry is an electrochemical technique based on the measurement of the potential between two electrodes without significant current flow.[18] The equipment required for potentiometric measurements is relatively simple and consists of an electrochemical cell composed of two electrodes, an indicator and a reference, immersed in a solution of the analyte, coupled with a potential-measuring device.[18,19] Briefly, the reference electrode is a half-cell with a known potential that is constant, regardless of the composition of the analyte solution, at a constant temperature. The indicator electrode varies its potential in a known manner depending on the activity of the analyte.[19]

Considering potentiometry as an analytical tool for quantitative purposes, the indicator electrode must respond selectively to the activity of the analyte of interest. Therefore, the ISE is widely used in potentiometry because it offers many advantages, such as fast response time, low-cost, wide linear range, flexibility in mounting in different shapes and sizes, and the ability to analyze colored and turbid samples.[18,20]

An ISE consists of a membrane, usually attached to the bottom of the tube containing the internal reference electrode. The reference solution is physically separated from the analyte solution by an ion-selective membrane. In particular, when developing ISEs for the determination of surfactants, the specific properties of these compounds are taken into account, such as the presence of hydrophobic moieties and the tendency to form ion pairs.[21] This membrane consists of an ion pair formed by the combination of a surfactant cation or other organic or inorganic cation and a surfactant anion (large organic or inorganic anion). It should have a low solubility product constant ($<10^{-12}$).[17] The membrane electrode and the external reference electrode are immersed in the solution containing the substance to be analyzed. Since the potential of the two reference electrodes is constant, any potential change in the cell is due to a potential change in the membrane. The ions in the solution are quantified by measuring the potential resulting from the distribution of the ions between the membrane and the solution.[17]

Polymeric membranes can be applied in different electrode designs (Figure 6.1). In one, the membrane separates the analyte solution from the

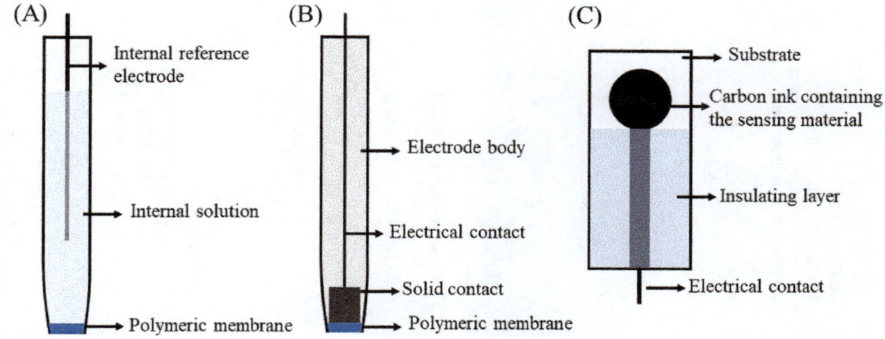

Figure 6.1 Schematic representation of the design of ion-selective electrodes: (A) liquid-contact ISE; (B) tubular solid-contact ISE, and (C) screen-printed ISE printed with homemade carbon ink.

interior of the conventional electrode body, which consists of an internal reference electrode (*e.g.*, silver/silver chloride) immersed in a suitable solution (Figure 6.1(A)). Another is the so-called solid-contact system, in which the polymeric membrane is in direct contact with a tubular or planar solid electrical conductor (Figure 6.1(B)). The solid-contact electrodes can be fabricated by applying the membrane to a solid electrode or by screen printing the electrodes (Figure 6.1(C)) with homemade carbon ink containing the electroactive compounds, polymer matrix, and plasticizer.[17,20,22]

With this background, the present review aims to briefly introduce different ISEs prepared with polymeric membranes containing surfactants in their composition, and their application in the determination of surfactants in different types of samples. The main properties and analytical characteristics of the surfactant-based potentiometric sensors for the determination of surfactants reported between 2014 and 2022 are summarized in Table 6.1.

The potentiometric ISEs presented in Table 6.1 were developed for primary use as indicator electrodes for the end-point in potentiometric titrations of ionic surfactants and their determination in various sample matrices. Basically, during titration, the ionic surfactant reacts with another surfactant of opposite charge to form a water-insoluble ion-pair complex. Therefore, the measured potential change before the equivalence point refers to the decreasing concentration of the surfactant under investigation. After the equivalence point, however, the entire amount of this surfactant is precipitated and the sensor potential responds to the increasing concentration of the titrant (oppositely charged surfactant). With this in mind, studies were conducted to determine experimental conditions that would not interfere with the titration curve and would allow accurate determination of the total surfactant content or the concentration of a particular surfactant. The following is a brief description of the main features of the ISEs developed for this purpose.

Table 6.1 Analytical characteristics of surfactant-based potentiometric sensors for surfactant determination in different sample types.[a]

Surfactant type: standard surfactant	Sensing material	Calibration range (M)	LOD (M)	Slope (mV per decade of activity)	Response time	pH range	Selectivity	Stability (lifetime)	Application	Ref.
Anionic: SDS SDBS	DDA-TPB	3×10^{-7}–8×10^{-3} 4×10^{-7}–2×10^{-3}	2×10^{-7}	59.0 59.1	Few seconds	2.0–12.0	Cl^-, I^-, F^-, CO_3^{2-}, HCO_3^-, NO_2^-, NO_3^-, SO_4^{2-}, HSO_4^-, PO_4^{3-}, $H_2PO_4^-$, CH_3COO^-, borate, benzoate, citrate, ethylenediamine tetraacetate, sulfamate, xylensulfonate, and toluensulfonate	>4 months with daily use	Powdered, gel, and liquid hand-dishwashing detergents	23
Anionic: SDS SDBS	DDA-TPB	Reported in ref. 23	2×10^{-7}	Reported in ref. 23	Not reported	2.0–10.0	Reported in ref. 23	>4 months with daily use	Industrial effluent samples	24
Anionic: SDS SDBS	DDA-TPB	In H_2O: 3.2×10^{-7}–4.6×10^{-3} 2.5×10^{-7}–1.2×10^{-3} In Na_2SO_4 0.01 M: 2.5×10^{-7}–4.6×10^{-3} 1.5×10^{-7}–8.8×10^{-4}	In H_2O: 2.5×10^{-7} 2.0×10^{-7} In Na_2SO_4 0.01 M: 2.5×10^{-7} 1.2×10^{-7}	In H_2O: 55.3 58.5 In Na_2SO_4 0.01 M: 55.4 57.7	In H_2O: 5 s (>1×10^{-4} M) 8 s (<1×10^{-5} M) In Na_2SO_4 0.01 M: 5 s (>1×10^{-4} M) 6 s (<1×10^{-5} M)	3.0–12.0	Cl^-, I^-, F^-, CO_3^{2-}, HCO_3^-, NO_2^-, NO_3^-, SO_4^{2-}, PO_4^{3-}, $H_2PO_4^-$, CH_3COO^-, borate, benzoate, citrate, ethylenediamine tetraacetate, sulfamate, xylene sulfonate, and toluene sulfonate.	~5 months with daily measurements	Gel and liquid hand-dishwashing detergents, industrial and household effluent samples	25
Anionic: SDS SDBS	DDA-TPB	7.2×10^{-7}–6.1×10^{-3} 6.8×10^{-6}–1.8×10^{-3}	3.9×10^{-7} 3.9×10^{-6}	−53.1 −48.1	Not reported	Not reported	HCO_3^-, HPO_4^{2-}, NO_3^-, SO_4^{2-}, HSO_4^-, CO_3^{2-}, $H_2PO_4^-$, CH_3COO^-, dodecylbenzene sulfonate, ethylenediamine tetraacetate, citrate, benzoate, toluene sulfonate, and xylensulfonate	Not reported	Commercial dishwashing detergents	26

Cationic: BTC	BT-TPB	5.97×10^{-7}–1.0×10^{-3}	3.92×10^{-7}	53.80	<10 s	2.0–10.0	Na^+, K^+, NH_4^+, Zn^{2+}, Mg^{2+}, Ba^{2+}, Pb^{2+}, Ca^{2+}, Co^{2+}, Cu^{2+}, and Fe^{3+}	1 month	BTC in pure form and pharmaceutical formulation of antibacterial spray; SDS, SLES, and LABSA in pure forms; SLES in shampoo, dishwashing liquid, and wastewater	27
Anionic: SDS	1-Dodecanthiol	6.0×10^{-6}–1.0×10^{-3}	5.0×10^{-6}	35.5	30 s (1×10^{-3} M)	3.0–5.0	SCN^-, ClO_4^-, IO_4^-, I^-, S^{2-}, CO^{2-}, CH_3COO^-, NO^-, PO_4^-, MoO_4^-, $S_2O_3^{2-}$, CN^-, Br^-, and Cl^-	~1 month or more	Mouth wash	28
Anionic: SDS STDS STTDS SHDS	[CuSalen] DS_2	4.0×10^{-7}–1.0×10^{-2} 5.0×10^{-7}–5.0×10^{-3} 5.0×10^{-7}–1.0×10^{-3} 5.0×10^{-7}–5.0×10^{-4}	2.0×10^{-7} 2.5×10^{-7} 2.5×10^{-7} 2.5×10^{-7}	56 59 62 64	7 s 7 s 8 s 9 s	3.0–12.0	DS-10, F^-, Cl^-, Br^-, SO_4^{2-}, NO_3^-, CH_3COO^-, and citrate	10–12 months	Water solutions, technical raw preparations, detergents, cosmetic and cleaners	21
Anionic: STDS STTDS SHDS	$[Cu(Dipyr)_2]$ DS_2	5.0×10^{-7}–5.0×10^{-3} 5.0×10^{-7}–1.0×10^{-3} 5.0×10^{-7}–5.0×10^{-4}	2.5×10^{-7}	56 61 64	8 s 8 s 9 s	3.0–10.0	NP-12, NO_3^-, CO_3^{2-}, SO_4^{2-}, F^-, Cl^-, Br^-, CH_3COO^-, and citrate	10–12 months	Water solutions, technical raw preparations, detergents, cosmetics, and cleaners	29

Table 6.1 (*Continued*)

Surfactant type: standard surfactant	Sensing material	Calibration range (M)	LOD (M)	Slope (mV per decade of activity)	Response time	pH range	Selectivity	Stability (lifetime)	Application	Ref.
Anionic: SDS SDBS	DDA-TPB	2×10^{-7}–7.0×10^{-3} 1.0×10^{-7}–2.0×10^{-3}	1.0×10^{-7} 1.0×10^{-7}	50.2 45.7	Not reported	Not reported	Cl^-, I^-, F^-, CO_2^-, HCO^-, NO_2^-, NO_3^-, SO_4^{2-}, HSO_4^-, PO_4^{3-}, $H_2PO_4^-$, CH_3COO^-, borate, benzoate, citrate, ethylenediamine tetraacetate, sulfamate, xylensulfonate, and toluensulfonate	60 measurements	Gel, powdered, and liquid hand-dishwashing detergents	30
Anionic: SDS SDBS	DDA-TPB	Not reported	Not reported	Not reported	20 s → (low concentrations of SDS) 60 s → (low concentrations of SDBS) 5 s → (higher concentrations)	2.0–10.0	Not reported	60 measurements	Wastewater samples	31
Anionic: SDS SDBS	DDA-TPB	In H_2O: 2.5×10^{-7}–4.5×10^{-3} 3.0×10^{-7}–2.5×10^{-3} In Na_2SO_4 0.01 M: 2.5×10^{-7}–4.5×10^{-3} 8.5×10^{-7}–4.5×10^{-3}	In H_2O: 1.5×10^{-7} 2.5×10^{-7} In Na_2SO_4 0.01 M: 1.5×10^{-7} 6.0×10^{-7}	In H_2O: 47.6 59.2 In Na_2SO_4 0.01 M: 51.2 62.1	12 s ($>1.0 \times 10^{-4}$ M) 15 s ($<1.0 \times 10^{-5}$ M)	2.0–7.0	Cl^-, I^-, F^-, CO_3^{2-}, HCO_3^-, NO_3^-, SO_4^{2-}, HSO_4^-, PO_4^{3-}, $H_2PO_4^-$, CH_3COO^-, borate, benzoate, citrate, ethylenediamine tetraacetate, sulfamate, xylensulfonate, toluensulfonate, dodecylbenzene sulfonate	>60 measurements per sensor	Gel and powdered laundry detergents, liquid hand-dishwashing detergent, and effluent samples	32

Cationic: CPC CTAB BTC	MWCNT-$OSO_3^-CP^+$	In H_2O: 1.5×10^{-7}–9.1×10^{-4} 3.5×10^{-7}–1.2×10^{-3} 2.0×10^{-6}–2.2×10^{-3} In Na_2SO_4 0.01 M: 1.6×10^{-7}–6.5×10^{-5} 4.5×10^{-7}–1.2×10^{-4} 2.5×10^{-6}–6.5×10^{-4}	In H_2O: 1.2×10^{-7} 2.5×10^{-7} 1.5×10^{-6} In Na_2SO_4 0.01 M: 1.2×10^{-7} 3.2×10^{-7} 1.5×10^{-6}	In H_2O: 58.9 56.6 52.1 In Na_2SO_4 0.01 M: 58.2 55.5 49.7	In H_2O: 5 s ($>5.0 \times 10^{-6}$ M) In Na_2SO_4 0.01 M: 9 s ($>5.0 \times 10^{-6}$ M)	3.0–10.0	NH_4^+, Na^+, Li^+, Mg^{2+}, Ca^{2+}, Zn^{2+}, ethanolamine, diethanolamine, triethanolamine, tetraethylammonium, tetramethyl ammonium, benzyltrimethyl ammonium, and benzyltriethylammonium	>6 months	Fabric softener, laundry disinfectant, and hand sanitizer	33
Anionic: SDS	DS-CP	5.8×10^{-7}–1.0×10^{-2}	5.8×10^{-7}	57.32	8 s	2.0–9.0	HCO_3^-, NO_2^-, Cl^-, CH_3COO^-, SO_4^{2-}, oxalate, SCN^-, benzoate, phthalate, NO_3^-, CO_3^{2-}, odecylbenzene sulfonate, and dioctylsulfo-succinate	5 months	Pharmaceutical preparations and spiked seawater and wastewater samples	34
Anionic: SDS	DS-CTA	6.3×10^{-7}–1.0×10^{-2}	6.3×10^{-7}	56.58	13 s	2.0–8.0	HCO_3^-, NO_2^-, Cl^-, CH_3COO^-, SO_4^{2-}, oxalate, SCN^-, benzoate, phthalate, NO_3^-, CO_3^{2-}, dodecylbenzene sulfonate, and dioctylsulfo-succinate	4–5 months	Pharmaceutical preparations and spiked seawater and wastewater samples	34
Anionic: STDS STTDS SHDS	$[Cu(Dipyr)_2]$ DS_2	5.0×10^{-7}–5.0×10^{-3} 5.0×10^{-7}–1.0×10^{-3} 5.0×10^{-7}–5.0×10^{-4}	2.5×10^{-7}	57 58 63	7 s 7 s 8 s	3.0–11.0	NP-12, NO_3^-, CO_3^{2-}, SO_4^{2-}, F^-, Cl^-, Br^-, CH_3COO^-, and citrate	5 months	Water solutions, technical → raw preparations, commercial detergents, cosmetic and cleaners	29

Table 6.1 *(Continued)*

Surfactant type: standard surfactant	Sensing material	Calibration range (M)	LOD (M)	Slope (mV per decade of activity)	Response time	pH range	Selectivity	Stability (lifetime)	Application	Ref.
Anionic: SDS STDS STTDS SHDS	[CuSalen] DS_2	4.0×10^{-7}–1.0×10^{-2} 4.0×10^{-7}–5.0×10^{-3} 5.0×10^{-7}–1.0×10^{-3} 5.0×10^{-7}–5.0×10^{-4}	2.0×10^{-7} 2×10^{-7} 2.5×10^{-7} 2.5×10^{-7}	57 57 61 62	6 s 6 s 7 s 7 s	4.0–10.0	Nonyl phenol-12, CH_3COO^-, NO_3^-, CO_3^{2-}, SO_4^{2-}, Cl^-, and Br^-	5 months	Cleaning product, preparation for producing detergents, synthetic detergent, and industrial wastewater	35
Cationic: TFACl	TFA-DS	8.2×10^{-7}–1.0×10^{-2}	8.2×10^{-7}	58.34	~10 s	2.0–8.5	Mg^{2+}, Ni^{2+}, Hg^{2+}, Cd^{2+}, Co^{2+}, Ca^{2+}, Fe^{2+}, Sr^{2+}, Na^+, K^+, Cl^-, SO_4^{2-}, Fe^{3+}, Al^{3+}, DTAB, Septonex, BMDS, potassium laurate, fatty acid ethoxylate, and sulfobetaines	~5 months	Water samples collected from different places	36

[a] Abbreviations – SDS: sodium dodecyl sulfate; SDBS: sodium dodecylbenzene sulfonate; DDA-TPB: dimethyldioctadecylammonium-tetraphenylborate; BTC: benzethonium chloride; BT-TPB: benzothonium-tetraphenylborate; SLES: sodium lauryl ether sulfate; LABSA: linear alkylbenzene sulfonic acid; STDS: sodium tridecylsulfate; STTD: sodium tetradecylsulfate; SHDS: sodium hexadecylsulfate; [CuSalen]: complex of copper(II) with *N*,*N*′-bis(salicylidene)ethylene diamine; [Cu(Dipyr)$_2$]DS$_2$: complex of copper(II) with α,α′-dipyridyl; MWCNT-$OSO_3^-CP^+$: MWCNTs modified with a sulfate group and cetylpyridinium ion; CPC: cetylpyridinium chloride; CTAB: hexadecyltrimethylammonium bromide; DS-CP: dodecylsulfate-cetypyridinium; DS-CTA: dodecylsulfate-cetyltrimethyl-ammonium; TFACl: *N*-(1,1,2,2-tetrahydroperfluorooctyl)-*N*,*N*-dimethylammonium chloride; FTA-DS: *N*-(1,1,2,2-tetrahydroperfluorooctyl)-*N*,*N*-dimethylammonium-dodecylsulfate; DTAB: dodecyltrimethylammonium bromide; BMDS: 1-butyl-3-methylimidazolium dodecylsulfonate.

6.2.1 Liquid-contact Ion-selective Electrodes

Various ion pairs synthesized with surfactants were prepared for the construction of ISEs for the determination of ionic surfactants in real samples.

Polymeric membranes based on the ion pair dimethyldioctadecylammonium-tetraphenylborate (DDA-TPB) have been extensively studied and used for surfactant determination. Galović *et al.* constructed a potentiometric sensor for the determination of anionic surfactants based on the ion pair DDA-TBP as the sensing material incorporated into a PVC membrane plasticized with nitrophenyl octyl ether (*o*-NPOE).[23] This sensor can be used in commercially available products with varying ionic strength and pH from 2.0 to 12.0 without significant effect on its performance or accuracy of determination. Some ethoxylated nonionic surfactants commonly found in anionic surfactant-based detergents were investigated as potential interferents and did not significantly interfere under typical real-world commercial product conditions. In addition, the potentiometric sensor for SDS proved to be highly selective toward various anions tested (Table 6.1). Potentiometric titration of anionic surfactants in three different commercial detergents was successfully performed using CPC as the titrant. The results were compared with those obtained by conventional two-phase titration and showed satisfactory agreement.

The same DDA-TPB ion pair was used for the determination of low concentrations of sodium dodecyl sulfate and sodium dodecylbenzene sulfonate in industrial effluent samples.[24] The potentiometric titrations of the industrial effluent samples (pH adjusted to 3.0) were performed with high accuracy and precision, and the results showed satisfactory correlation compared with those obtained by the standard extraction spectrophotometric MBAS method.

Later, Samardžić *et al.*[25] added 10% (w/w) carbon powder to the composition of the polymeric membrane with the ion pair DDA-TPB reported by Galović *et al.*[23] The authors affirmed that the presence of the carbon powder in the membrane resulted in a reduction of ohmic resistance by more than three orders of magnitude, which led to a reduction in noise and signal drift. The electrical and analytical properties of the proposed sensor were also significantly improved. Potentiometric titration using the developed sensor as the end-point indicator and CPC as the titrant was effective in determining the content of anionic surfactants in two different detergents and five effluent samples.

Karnaš *et al.* focused their study on evaluating the influence of different plasticizers on the properties of the PVC-based membrane sensor containing the ion pair DDA-TPB for determining the content of anionic surfactants in commercial products.[26] Six different plasticizers (2-nitrophenyl octyl ether, bis(2-ethylhexyl) phthalate, bis(2-ethylhexyl) sebacate, 2-nitrophenyl phenyl ether, dibutyl phthalate, and dibutyl sebacate) were incorporated into the membrane formulations, which were characterized by direct potentiometric response and dynamic response. According to the authors, the PVC membrane plasticized with dibutyl phthalate showed the best analytical performance, exhibiting a fast dynamic response and the highest value of potential change at the end-point, allowing reliable titration of the analyte, even at low concentrations. In addition, dibutyl phthalate is a low-cost substance compared to the other commonly used plasticizers. Therefore, the developed sensor was also tested in the presence of other anions (Table 6.1) and no significant effect on the potentiometric measurements was observed. Finally, the content of anionic surfactants in commercial dishwashing detergents was successfully determined using the DDA-TPB sensor containing the plasticizer dibutyl phthalate.

The polymeric membrane sensor based on the incorporation of the ion pair benzothonium-tetraphenylborate (BT-TPB) was constructed for the determination of the cationic surfactant BTC and some other anionic surfactants.[27] The authors studied the composition of the membrane in terms of the amount of the ion pair BT-TPB, considering a fixed ratio of the plasticizer tricresyl phosphate to PVC polymer (1 : 1). The best analytical performance in terms of slope, linear range, and limit of detection was obtained with the membrane containing 5% (w/w) BT-TPB. The proposed sensor showed high selectivity for BT ions over various inorganic cations tested (Table 6.1). Therefore, it was used as an end-point indicator electrode in a potentiometric titration of BT ion in its pure form and pharmaceutical formulation of an antibacterial spray using TPB as titrant. In addition, the developed sensor was also used for the determination of anionic surfactants, SDS, sodium dodecyl ether sulfate and linear alkylbenzene sulfonic acid in its pure form, and sodium dodecyl ether sulfate in the samples of shampoo, dishwashing liquid, and wastewater using BTC as titrant.

Abounassif *et al.* used the surfactant 1-dodecanthiol as a sensing material in the construction of a potentiometric membrane-based sensor for the determination of SDS.[28] The authors evaluated dibutyl sebacate, dioctyl phthalate, and *o*-NPOE as plasticizers for the formulation of the active membrane based on a PCV polymer matrix. According to the results, dioctyl phthalate proved to be the most suitable plasticizer, improving the solubility and selectivity of the membrane, and providing the optimal mechanical properties of the membrane, unlike the others. Studies indicated moderate selectivity of the sensor for SDS in the presence of many anions. SDS was determined in mouth wash by potentiometric titration using this sensor as the end-point indicator and by the standard spectrophotometric method. The results agreed well.

6.2.2 Solid-contact Ion-selective Electrodes

Electrodes in which the active membrane is in direct contact with a conductive solid are less fragile and more robust, durable, and can be easily miniaturized.[20,37]

Polymeric membranes containing ion associates of SDS and cationic complexes of copper(II) with organic reagents (1,10-phenanthroline,

N,*N*′-bis(salicylidene)ethylene diamine, α,α′-dipyridyl) were used as electroactive compounds for the determination of homologous sodium alkyl sulfates (SDS, sodium tridecylsulfate, sodium tetradecylsulfate, and sodium hexadecylsulfate) in different samples.[21,29] PVC membranes plasticized with dibutyl phthalate were used in solid-contact sensors with the electrical conductor graphite. The sensors showed stable electrochemical and operational properties as well as fast response time, nearly Nernstian slope, high sensitivity, and good selectivity for anionic surfactants. The authors of both studies showed the results obtained in the determination of anionic surfactants in raw materials, detergents, cosmetic, and cleaning agents using sensors based on the electroactive compound $[CuSalen]DS_2$[21] and $[Cu(Dipyr)_2]DS_2$[29] by potentiometric titration with CPC as titrant.

Potentiometric microsensors based on the application of an appropriate volume of membrane solution to the surface of a commercial carbon-based screen-printed electrode were used for the determination of anionic surfactants in commercial detergents[30] and wastewater samples.[31] In both works, the authors used the previously reported ion pair DDA-TPB[23] as the sensing material, PVC as the polymer matrix, and *o*-NPOE as the plasticizer. All analyzed samples were titrated using the developed microsensors as the end-point indicator electrode and CPC as the titrant. The results showed satisfactory agreement with those obtained by standard methods (two-phase titration or MBAS) and with conventional surfactant ISEs based on the same sensing material.

Later, a new proposal for a microsensor with the ion pair DDA-TPB was reported by Galović *et al.*[32] It was based on immobilization of the sensing material in the polymeric membrane with graphene. The membrane was prepared according to the previously reported method,[23] but with the addition of 1% graphene nanoparticles to the composition. A suitable volume of the membrane solution was dropped onto the screen-printed carbon-based electrode and dried for 24 h. The incorporation of graphene into the membrane resulted in a significant reduction in electrical resistance and thus signal noise, and also prevented leaching of the sensing material from the membrane. The sensor showed excellent selectivity performance with slight interference from toluene sulfonate, xylene sulfonate, and hydrogen carbonate. Nevertheless, it was successfully used in the determination of anionic surfactants (SDS and dodecylbenzenesulfonate) in commercial products and effluent samples.

Hajduković *et al.* prepared a hybrid sensing material for the determination of cationic surfactant using multi-walled carbon nanotubes modified with a sulfate group and a cetylpyridinium ion.[33] This material was mixed with PVC, *o*-NPOE, tetrahydrofuran, and dimethylformamide. It was then dropped onto the graphite electrode surface coated with graphite ink (solid-contact sensor). The authors reported that the hybrid sensing material showed similar behavior to a solid-state electrode with a liquid membrane and also exhibited greater membrane stability, negligible noise and signal drift, significantly less leaching of the sensing material from the membrane, and thus longer sensor lifetime. The developed sensor exhibited exceptional selectivity for cetylpyridinium over all organic and inorganic cations tested and was used as an end-point indicator electrode in the potentiometric titration of cationic surfactants in various commercial products. The results showed no significant differences from the results of the two-phase titration.

Awad Ali and Mohamed developed ISEs based on the use of ion association complexes of SDS with CPC and CTA as exchange sites in a screen-printed electrode matrix for the determination of SDS in various samples.[34] The sensors were printed with a homemade carbon ink prepared by mixing carbon powder, the ion pair as sensing material, PVC as the polymeric matrix, and dibutyl phthalate as the plasticizer. The selectivity of the two ISEs was evaluated and, with the exception of the dodecylbenzenesulfonate anion, there was no significant interference from most of the tested substances. Moreover, better results in terms of selectivity were obtained with the ion pair DS-CP. Both ISEs were successfully used as the end-point indicator electrode in potentiometric titrations of SDS in pharmaceutical preparations and in real spiked water samples.

The ionic associations of SDS and cationic complexes of copper(II) with organic reagents 1,10-phenanthroline, *N*,*N*′-bis(salicylidene)ethylene diamine, and α,α′-dipyridyl were also explored as electroactive compounds for the development of homemade potentiometric screen-printed sensors.[29,35] The homemade carbon ink was prepared with carbon nanotubes. The electroanalytical properties of the planar electrodes were compared with those of tubular ISEs with membranes based on the ion pairs studied. The sensors were successfully used for the determination of homologous sodium alkyl sulfates in different samples.

Ali *et al.* used *N*-(1,1,2,2-tetrahydroperfluorooctyl)-*N*,*N*-dimethylammonium chloride and SDS as exchange sites in a screen-printed electrode.[36] The working screen-printed electrode was printed with homemade carbon ink. No significant interference was observed from the tested inorganic cations and anions. Meanwhile, the surfactant dodecyltrimethylammonium bromide seriously affected the sensor performance and septonex showed slight interference with *N*-(1,1,2,2-tetrahydro perfluorooctyl)-*N*,*N*-dimethylammonium chloride. The ISE was tested *via* its application for the determination of *N*-(1,1,2,2-tetrahydro perfluorooctyl)-*N*,*N*-dimethylammonium chloride in different water samples collected from different places. Good agreement was obtained between the results obtained by potentiometric titration using this ISE as the end-point indicator electrode and by a standard two-phase titration method.

6.3 Conclusion

This brief overview makes it clear that there are ISEs designed with polymeric membranes immobilized on liquid-contact and also on solid-contact for the determination of surfactants in various complex sample types. ISEs developed with solid-contact are characterized by robustness and ease of miniaturization. In addition, the polymeric membrane based on the ion pair dimethyldioctadecylammonium tetraphenylborate (DDA-TPB) has been most commonly used for the development of ISEs with liquid- or solid-contacts. Finally, it was found that the vast majority of developed ISEs were used in the determination of anionic surfactants, such as SDS, by potentiometric titration.

Acknowledgements

The authors gratefully acknowledge financial support and scholarships from the Brazilian funding agencies CNPq [grant numbers 408591/2018-8, 305320/2019-0], Fundação Araucária do Paraná, and CAPES [grant number 88887.674846/2022-00].

References

1. S. A. Mousavi and F. Khodadoost, Effects of detergents on natural ecosystems and wastewater treatment processes: a review, *Environ. Sci. Pollut. Res.*, 2019, **26**(26), 26439–26448.
2. T. Ivanković and J. Hrenović, Surfactants in the Environment, *Arch. Ind. Hyg. Toxicol.*, 2010, **61**(1), 95–110.
3. G.-G. Ying, B. Williams and R. Kookana, Environmental fate of alkylphenols and alkylphenol ethoxylates—a review, *Environ. Int.*, 2002, **28**(3), 215–226.
4. K. Jardak, P. Drogui and R. Daghrir, Surfactants in aquatic and terrestrial environment: occurrence, behavior, and treatment processes, *Environ. Sci. Pollut. Res.*, 2016, **23**(4), 3195–3216.
5. B. Bina, F. Mohammadi, M. M. Amin, H. R. Pourzamani and Z. Yavari, Determination of 4-nonylphenol and 4-*tert*-octylphenol compounds in various types of wastewater and their removal rates in different treatment processes in nine wastewater treatment plants of Iran, *Chin. J. Chem. Eng.*, 2018, **26**(1), 183–190.
6. M. Palmer and H. Hatley, The role of surfactants in wastewater treatment: Impact, removal and future techniques: A critical review, *Water Res.*, 2018, **147**, 60–72.
7. N. Alygizakis, A. Galani, N. I. Rousis, R. Aalizadeh, M.-A. Dimopoulos and N. S. Thomaidis, Change in the chemical content of untreated wastewater of Athens, Greece under COVID-19 pandemic, *Sci. Total Environ.*, 2021, **799**, 149230.
8. A. A. Siyal, M. R. Shamsuddin, A. Low and N. E. Rabat, A review on recent developments in the adsorption of surfactants from wastewater, *J. Environ. Manage.*, 2020, **254**, 109797.
9. CESIO Industry Statistics 2021. Available from: https://www.cesio.eu/index.php/information-centre/industry-data.
10. H. Sun, S. Zhou, Y. Jiang, Z. Qin, J. Fei and Y. Sun, *et al.*, Effect of cationic, anionic and non-ionic surfactants on transport of microplastics:

Role of adhesion of surfactants on the polyethylene surface, *J. Hydrol.*, 2022, **612**, 128051.
11. E. Jurado, M. Fernández-Serrano, J. Núñez-Olea, G. Luzón and M. Lechuga, Simplified spectrophotometric method using methylene blue for determining anionic surfactants: Applications to the study of primary biodegradation in aerobic screening tests, *Chemosphere*, 2006, **65**(2), 278–285.
12. B. Wyrwas and A. Zgoła-Grześkowiak, Continuous Flow Methylene Blue Active Substances Method for the Determination of Anionic Surfactants in River Water and Biodegradation Test Samples, *J. Surfactants Deterg.*, 2014, **17**(1), 191–198.
13. M. Tsubochi, N. Yamasaki and K. Matsuoka, Determination of anionic surfactants by two-phase titration with tetrabromophenolphthalein ethyl ester as indicator, *J. Am. Oil Chem. Soc.*, 1979, **56**(11Part2), 921–923.
14. N. Sakač, D. Madunić-Čačić, D. Marković, L. Hok, R. Vianello and B. Šarkanj, *et al.*, Potentiometric Surfactant Sensor Based on 1,3-Dihexadecyl-1*H*-benzo[*d*]imidazol-3-ium for Anionic Surfactants in Detergents and Household Care Products, *Molecules*, 2021, **26**(12), 3627.
15. C. Miller, B. S. Bageri, T. Zeng, S. Patil and K. K. Mohanty, Modified Two-Phase Titration Methods to Quantify Surfactant Concentrations in Chemical-Enhanced Oil Recovery Applications, *J. Surfactants Deterg.*, 2020, **23**(6), 1159–1167.
16. J. Sánchez and M. del Valle, Determination of Anionic Surfactants Employing Potentiometric Sensors—A Review, *Crit. Rev. Anal. Chem.*, 2005, **35**(1), 15–29.
17. M. Jozanović, N. Sakač, M. Karnaš and M. Medvidović-Kosanović, Potentiometric Sensors for the Determination of Anionic Surfactants – A Review, *Crit. Rev. Anal. Chem.*, 2021, **51**(2), 115–137.
18. J. Wang, Potentiometry, in *Analytical Electochemistry*, John Wiley & Sons, Inc., New York, USA, 2nd edn, 2002, pp. 140–170.
19. D. A. Skoog, D. M. West, F. J. Holler and S. R. Crouch, *Fundamentals of analytical chemistry*, Thomson-Brooks/Cole, Belmont, CA, 8th edn, 2004.
20. L. Tymecki, S. Glab and R. Koncki, Miniaturized, Planar Ion-selective Electrodes Fabricated by Means of Thick-film Technology, *Sensors*, 2006, **6**(4), 390–396.
21. N. M. Makarova and E. G. Kulapina, New Potentiometric Sensors Based on Ionic Associates of Sodium Dodecylsulfate and Cationic Complexes of Copper(II) with some Organic Reagents, *Electroanalysis*, 2015, **27**(3), 621–628.
22. Y. Shao, Y. Ying and J. Ping, Recent advances in solid-contact ion-selective electrodes: functional materials, transduction mechanisms, and development trends, *Chem. Soc. Rev.*, 2020, **49**(13), 4405–4465.
23. O. Galović, M. Samardžić, S. Petrušić and M. Sak-Bosnar, A New Sensing Material for the Potentiometric Determination of Anionic Surfactants in Commercial Products, *Int. J. Electrochem. Sci.*, 2014, **9**, 3802–3818.
24. M. Samardžić, O. Galović, S. Petrušić and M. Sak-Bosnar, The Analysis of Anionic Surfactants in Effluents Using a DDA-TPB Potentiometric Sensor, *Int. J. Electrochem. Sci.*, 2014, **9**, 6166–6181.

25. M. Samardžić, O. Galović, M. Hajduković and M. Sak-Bosnar, A novel, fast responding, low noise potentiometric sensor containing a carbon-based polymeric membrane for measuring surfactants in industrial and environmental applications, *Talanta*, 2017, **162**, 316–323.
26. M. Karnaš, N. Sakač, M. Jozanović, M. Tsakiri, M. Kopriva and E. K. Andrić, *et al.*, The Influence of Plasticisers on Response Characteristics of Anionic Surfactant Potentiometric Sensor, *Int. J. Electrochem. Sci.*, 2017, **12**, 5921–5933.
27. Y. M. Issa, S. H. Mohamed and M. A.-E. Baset, Chemically modified carbon paste and membrane sensors for the determination of benzethonium chloride and some anionic surfactants (SLES, SDS, and LABSA): Characterization using SEM and AFM, *Talanta*, 2016, **155**, 158–167.
28. M. A. Abounassif, M. M. Hefnawy, H. Al-Robian and G. A. E. Mostafa, Dodecanthiol as Novel Sensing Material for Potentiometric Determination of Sodium Dodecyl Sulphate Anionic Surfactant, *Int. J. Electrochem. Sci.*, 2015, **10**, 8668–8679.
29. N. M. Makarova and E. G. Kulapina, New potentiometric screen-printed sensors for determination of homologous sodium alkylsulfates, *Sens. Actuators, B*, 2015, **210**, 817–824.
30. O. Galović, M. Samardžić and M. Sak-Bosnar, A New Microsensor for the Determination of Anionic Surfactants in Commercial Products, *Int. J. Electrochem. Sci.*, 2015, **10**, 5176–5193.
31. O. Galović, M. Samardžić, S. Petrušić and M. Sak-Bosnar, Application of a New Potentiometric Sensor for Determination of Anionic Surfactants in Wastewater, *Chem. Biochem. Eng. Q.*, 2015, **29**(3), 307–313.
32. O. Galović, M. Samardžić, M. Hajduković and M. Sak-Bosnar, A new graphene-based surfactant sensor for the determination of anionic surfactants in real samples, *Sens. Actuators, B*, 2016, **236**, 257–267.
33. M. Hajduković, M. Samardžić, O. Galović, A. Széchenyi and M. Sak-Bosnar, A functionalized nanomaterial based, new, solid state cationic-surfactant-selective sensor with fast response and low noise, *Sens. Actuators, B*, 2017, **251**, 795–803.
34. T. A. Ali and G. G. Mohamed, Modified Screen-Printed Ion Selective Electrodes for Potentiometric Determination of Sodium Dodecylsulfate in Different Samples, *J. AOAC Int.*, 2015, **98**(1), 116–123.
35. N. M. Makarova and E. G. Kulapina, Planar electrodes based on carbon nanotubes for the potentiometric determination of homologous sodium alkyl sulfates, *J. Anal. Chem.*, 2015, **70**(7), 879–884.
36. T. A. Ali, M. H. Soliman and G. G. Mohamed, Electrochemical Determination of *N*-(1,1,2,2-tetrahydroperfluorooctyl)-*N*,*N*-dimethylammonium Chloride Surfactant in Different Water Samples Using Modified Screen-Printed Electrode, *Int. J. Electrochem. Sci.*, 2016, **11**, 1055–1069.
37. Y. H. Cheong, L. Ge and G. Lisak, Highly reproducible solid contact ion selective electrodes: Emerging opportunities for potentiometry – A review, *Anal. Chim. Acta*, 2021, **1162**, 338304.

CHAPTER 7

Surfactant Sensors for Environmental Analysis

SWETAPADMA PRAHARAJ AND DIBYARANJAN ROUT*

Department of Physics, School of Applied Sciences, KIIT Deemed to be University, Bhubaneswar-24, Odisha, India
*Email: droutfpy@kiit.ac.in

7.1 Introduction

Rapid industrialization and population growth have resulted in massive environmental contamination all around the globe. Among the various types of pollution, water pollution has taken a front foot in affecting the lives of not only humans but also aquatic plants and animals. Various effluents in the form of dyes, heavy metals, polymers, organic compounds *etc.* from industries and households (*e.g.* food, cosmetics, textiles, rubber, drugs and pharmaceuticals) contaminating water bodies become more problematic when they are not degraded by conventional methods. Some of these emerging pollutants are surfactants, which need utmost attention in water treatment processes. Surfactants, otherwise known as surface active agents, are usually amphiphilic entities having both hydrophilic and lipophilic/hydrophobic groups. The most important function of a surfactant is to modulate the energy relationship at interfaces in terms of alteration of interfacial or surface tension. Addition of surfactants in small concentrations lessens the amount of interfacial energy by promoting the expansion of one of the surfaces at the cost of other. This improves the efficiency of numerous processes, thereby making it a vital component of detergents, wetting agents, emulsifying agents, disinfectants, cleansers, conditioners, *etc.*

Detection Science Series No. 23
Surfactant-based Sensors in Chemical and Biochemical Detection
Edited by J. G. Manjunatha

Published by the Royal Society of Chemistry, www.rsc.org

A schematic diagram depicting the fundamental structure of a surfactant is given in Figure 7.1.

A survey of the global surfactant market in the year 2021 estimated it to be worth USD 40.5 billion and this is anticipated to climb to USD 53.3 billion by the end of 2028. At the same time, there are increasing global environmental concerns, especially with regard to waste water. The harmful effects of surfactants on the environment, wildlife and humans has been studied vividly in the past by many researchers.[1,2] In a report by Venhuis and Mehrvar,[1] a popular surfactant, linear alkylbenzene sulphonate (LAS) was found to harm the fish gills and disrupt swimming patterns in the larvae of blue mussels. Not only that, LAS also causes excess secretion of mucus, which reduces respiration in the common goby. In a separate study, it was noticed that surfactants tend to retain an amount of hydrocarbons in water resources and soil, which can significantly harm crops.[3] It has also been seen that traces of surfactant (like benzalkonium chloride) can increase antibiotic resistance in microbes.[4] Though several techniques have been employed in the past to remove these surfactants, but no method is efficient enough for 100% removal. As a result they can bioaccumulate within certain species and lead to tissue mutations. Therefore it is necessary to assess the amount of harmful surfactants in different sources before they are put into real time use. In this regard, different sensing methods for quantifying the amount of surfactant pollutant in the environmental samples are used. On the other hand, environmentally benign surfactants may act as useful agents in voltametric method to detect pollutants in various sources. With these considerations, the current chapter focuses on surfactant sensing mechanisms and techniques. Additionally, the usefulness of surfactant-based voltametric sensors for detection of environmental pollutants is also discussed.

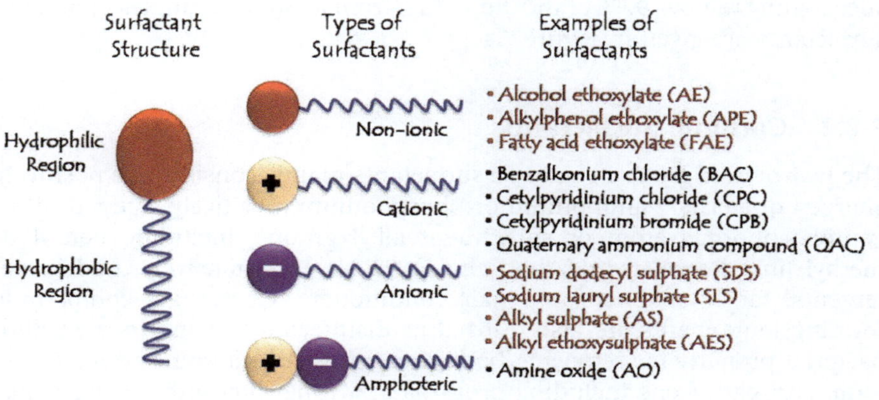

Figure 7.1 Schematic figure showing the basic structure of a surfactant and classification of surfactants with examples.

7.2 Types of Surfactant

The current surfactant treatment methods are facing challenges due to the variety of surfactants available in the market with diverse characteristics. Within this domain, the main types of surfactant are anionic, cationic, amphoteric and non-ionic (Figure 7.1).

7.2.1 Anionic Surfactants

Anionics constitute the oldest and most familiar category of surfactants often found in soaps and detergents, *etc.* They usually contain a negatively charged hydrophilic group like carboxyl, sulphate, sulphonate or phosphate, while the hydrophobic part is normally an alkyl chain of varying length, an alkylbenzene or an alkylphenyl ether. Apart from their use in soaps, anionic surfactants have been applied in the biotechnological and cosmetic industries. In the pharmaceutical industry, these surfactants play a decisive role in the formulation of drug delivery systems. Especially, micellar drug delivery systems involving organosulphate or sodium dodecyl sulphate (SDS) are attractive due to the ease of availability, cost effectiveness and high foaming properties. A molecule of SDS comprises a 12-carbon tail anchored onto a sulphate group and is amphiphilic in nature. It has inherent skin cleansing and antimicrobial properties and is well known for its rigorous interactions owing to the oxygen groups, which act as donors as well as acceptors.[5] Moreover, surfactants can affect the function of proteins by altering their structure, thereby denaturing them at low concentrations. They are also known to impact the aggregation of protein.[6] In addition, anionic surfactants are also useful in removing petrochemical products from soil. In relation to that, Khalladi *et al.*[7] investigated the effect of sodium dodecyl sulphate (SDS) on the removal of petroleum hydrocarbons from contaminated soil. They observed that washing of the soil at 3.2 $ml\,min^{-1}$ could successfully remove 97% of the diesel fuel from it. Some examples of anionic surfactants are given in Figure 7.1.

7.2.2 Cationic Surfactants

The hydrophilic group in cationic surfactants mainly consists of a positively charged quaternary ammonium or phosphonium (positively charged nitrogen/phosphorous atom or substituent alkyl groups, including benzyl or methyl linked to a hydrocarbon chain). These are extensively used in detergents, fabric softeners and hair conditioners. Quaternary compounds forming long chains are also utilized as disinfectants owing to their antibacterial property for removing both gram positive and gram negative bacteria, and pathogens including protozoa and fungi. Recently, Wang *et al.*[8] took advantage of the synergy between a cationic surfactant pi-functionalized quaternary ammonium (QA) compound and graphene oxide to remove

Gram negative *Escherichia coli* (*E. coli*) and Gram positive *Staphylococcus aureus* (*S. aureus*) from water. The large aromatic cores of piQA allowed the formation of collectively strong pi–pi interactions with reduced graphene oxide leading to high piQA mass loading, ∼31 wt%. This helped the high antibacterial activity of the combination. Moreover, the piQA assembly was less toxic for the two different mammalian cell lines, *i.e.* L929 and macrophages. However, there are previous reports on the retarding effects of cationic surfactants on human lymphocytes and rabbit epithelial cells. Apart from that, this category of surfactants has also been instrumental in reducing the size of nanoparticles. In 2011, Singh *et al.*[9] studied the effect of a series of cationic surfactants such as dodecyltrimethylammonium bromide (DTAB), tetradecyltrimethylammonium bromide (TTAB) and cetyltriammonium bromide (CTAB) of varying chain lengths on the grain size of silica nanoparticles. The effect of different surfactants on the surface morphology and grain size of different surfactants are shown in Figure 7.2. An increase in the chain length resulted in a decrease in the size of the nanoparticles. Further, the incorporation of reduced-size silica particles into cement paste enhanced its mechanical properties. Some examples of cationic surfactants are given in Figure 7.1.

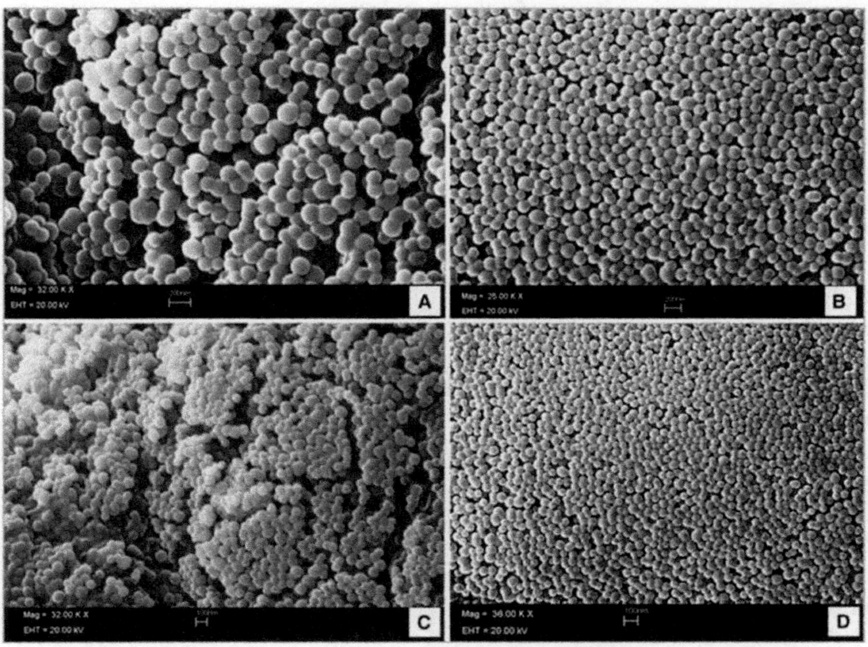

Figure 7.2 Surface morphology of silica nanoparticles prepared (A) without surfactant (∼180 nm), with (B) DTAB (∼140 nm), (C) TTAB (∼95 nm) and (D) CTAB (∼55 nm). Reproduced from ref. 9 with permission from Springer Nature, Copyright 2011.

7.2.3 Amphoteric Surfactants

Surfactants possessing a hydrophobic long hydrocarbon chain along with positive and negative charged hydrophilic centres connected to each other by a spacer group are said to be amphoteric. Due to the existence of both positively and negatively charged elements, the amphoteric surfactants maintain overall charge neutrality and their properties primarily depend upon the hydrocarbon chain length, number of spacer methylene groups and the relative positions of the charged (positive/negative) groups. Moreover, the ionic response of this group of compounds is controlled by the pH of the solvent, *i.e.* cationic nature is exhibited below the isoelectric points while anionic behaviour is observed at higher pH values.

These molecules can even take a zwitterionic form around the isoelectric point. A combination of such a unique molecular configuration leads to exotic properties such as high aqueous solubility, surface activity, large isoelectric range, good foam stability, less toxicity, low critical micellar concentration, outstanding bioactivity and degradability, *etc.* With a permutation of properties, amphoteric surfactants are widely applied in the field of chromatography, cosmetics, electrochemistry, waste water treatment, improved oil recovery, *etc.* Aqueous solutions of some of these surfactants give exceptionally low surface tension compared to water, which results in good washing capability. For example, the interaction between surfactant and hair is mainly physical rather than being ionic and hence it can be rinsed quickly from the hair surface. Moreover, the biocompatibility of amphoteric surfactants like cocamidopropyl betaine (CAPB) also reduces irritation of the skin compared to anionic sodium laureth sulphate (SLES) and sodium lauryl sulphate (SDS).[10] Recently, Zieba *et al.*[11] fabricated some new sulphobetaines and betaines demonstrating good washing efficiency with low risk of skin and eye irritation. In addition, long hydrocarbon chains (~C12) with hydrophobicity are capable of exhibiting enhanced oil recovery under harsh conditions. Conventionally, amphoteric surfactants based on carboxybetaine generate ultralow interfacial tension at low critical micelle concentration (CMC).[12] On the other hand, ethoxy groups included in the hydrophobic chain of sulphobetaine are found to improve water solubility and lessen the necessity for co-solvent for low microemulsion viscosity.[13] Similarly, in the treatment of waste water, amphoteric surfactants can hinder the water molecules vigorously in the area of amphoteric ionic heads, thus producing a tight hydration shell to filter out impurities in water.[14] In 2013, Bergaya *et al.*[15] observed good cation exchange capability and large specific surface area of montmorillonite (naturally abundant swelling clay used in environmental remediation). Examples of aamphoteric surfactants are given in Figure 7.1.

7.2.4 Non-ionic Surfactants

Non-ionic surfactants obtain their surface activity from equilibrium between hydrophobic and hydrophilic elements in the structure. Usually, this category of surfactant cannot dissociate into ions in aqueous media. However, their solubility is decided by the constituent polar head groups. While the hydrophilic group can be different chain lengths of ethylene oxide, the hydrophobic part may consist of a fatty acid, alkylated phenol derivative or linear alcohol having a long chain. With these unique properties, non-ionic surfactants have been frequently employed as wetting agents, foam stabilization agents and emulsifiers. Apart from that, these compounds can efficiently improve drug carrier stability and also form a major part of pesticide formulations. In some cases, it has been observed that a combination of surfactants rather than one surfactant is more effective in different applications than a single component. In this regard, Li *et al.*[16] found that a combination of two non-ionic surfactants in the preconcentrate could efficiently produce self-emulsifying microemulsions with small particle size. Further, the two component system increased the drug loading stability of the emulsion. It is expected to have implications in better management of poorly aqueous soluble drugs for self-emulsifying microemulsion drug delivery systems. Some examples of non-ionic surfactants are given in Figure 7.1.

7.3 Surfactant Sensors

Owing to the toxic effect of some of surfactants on the environment, different types of sensors, including potentiometric, voltammetric, amperometric and even biosensors are used. While potentometric sensors are the most commonly employed due to their versatility and simplicity, amperometric sensors and biosensors are rarely used for sensing surfactants. On the other hand surfactant-based sensors have also become popular for the detection of environment pollutants such as heavy metals and gases. In this section, we will discuss both of them separately.

7.3.1 Sensors for Surfactant Determination

7.3.1.1 Surfactant Sensors Based on a Potentiometric Method

A potentiometric sensor is frequently referred to as a titration end-point sensor in batch mode, though it can be employed continuously in FIA (flow injection analysis) mode. These sensors can be utilized in diverse and extreme experimental conditions, making them potentially useful for environmental monitoring and industrial process control. Recently, potentiometric sensors have been employed for surfactant sensing with no pretreatment of samples (for coloured and viscous samples also) and simplicity of measurement. These sensors are based on the relation between the activity of the analyte and the potential response. The first of this kind of sensor were used to detect anionic surfactants, appearing in the 1960s. Moreover the liquid membrane electrodes were effective in analyzing the surfactant since they comprised an ion exchange solution almost similar to ion pairs in the surfactant solution. Sap *et al.*[17] employed the Fe(II) bis[2,4,6-tri(2-pyridyl)]-1,3,5-triazine complex, which was implanted into the sensor to determine the amount of dodecyl sulphate (DS). This sensor based on a liquid membrane exhibited linear response in the range 7×10^{-6} to 1.5×10^{-6} mol L^{-1}. Further, a selective liquid in combination with a PVC-based membrane for the determination of dodecyl sulphate was developed by Arvand-Barmchia *et al.*[18] In 2016, Galovic *et al.*[19] designed a highly sensitive potentiometric all solid-state sensor using dimethyldioctadecyl-ammonium-tetraphenylborate (DDA-TPB) ion pair for sensing surfactants, especially anionic. The sensor exhibited Nernstian reactions for DBS (dodecylbenzenesulphate), with sub-Nernstian behaviour for dodecyl sulphate (DS) from 2.5×10^{-7} to 4.5×10^{-3}. Figure 7.3 summarizes the potentiometric results obtained for the above-mentioned sensor.

7.3.1.2 Sensors Based on Voltammetry and Amperometric Methods

Usually, surfactants appear to be non-electroactive to voltammetric and amperometric methods due to which it becomes essential to employ an indirect approach, *i.e.* applying a solution with specific amounts of electroactive species (normally, a marker). The solution with the surfactant is reduced/oxidized close to the working electrode. Upon using a marker, the current produced in it reduces in the presence of the surfactant which again leads to inhibition of electron transfer at the electrode. This reduction in the current is usually related to the concentration of the analyte. This is one of the major drawbacks in this and was investigated by Skoog *et al.*[20] They studied the influence of some surfactants on the amperometric and cyclic voltammetric response of hexacynoferrate. They observed an irreversible blocking effect on the electrode response due to the presence of non-ionic surfactants. Such an effect was later studied by Adeloju and Shaw,[21,22] employing different types of working electrodes such as the hanging mercury drop electrode (HMDE), a mercury film electrode or a dropping mercury electrode. In addition, an interesting work on a bacteria-based amperometric biosensor for detection of anionic surfactants was carried out by Taranova *et al.*[23] in 2002. They tested several strains of bacteria out of which sensors based on *Pseudomonas* and *Achromobacter* genera exhibited high sensitivity towards sodium alkansulphonate and some non-ionic surfactants (Figure 7.4). These microbial amperometric sensors enabled detection of surfactants with high selectivity, sensitivity and reproducibility of signals for prolonged periods of time. Further, an amalgamation of differential pulse polarography with ion-pair extraction was found to be appropriate for the determination of 10^{-7} mol L^{-1} of organic quaternary ammonium bases applying Orange II as a suitable polarographic active counter ion. This technique was typically sued for detecting surfactants like tetrapentylammonium bromide (TPAB), and codeine and septonex.[24]

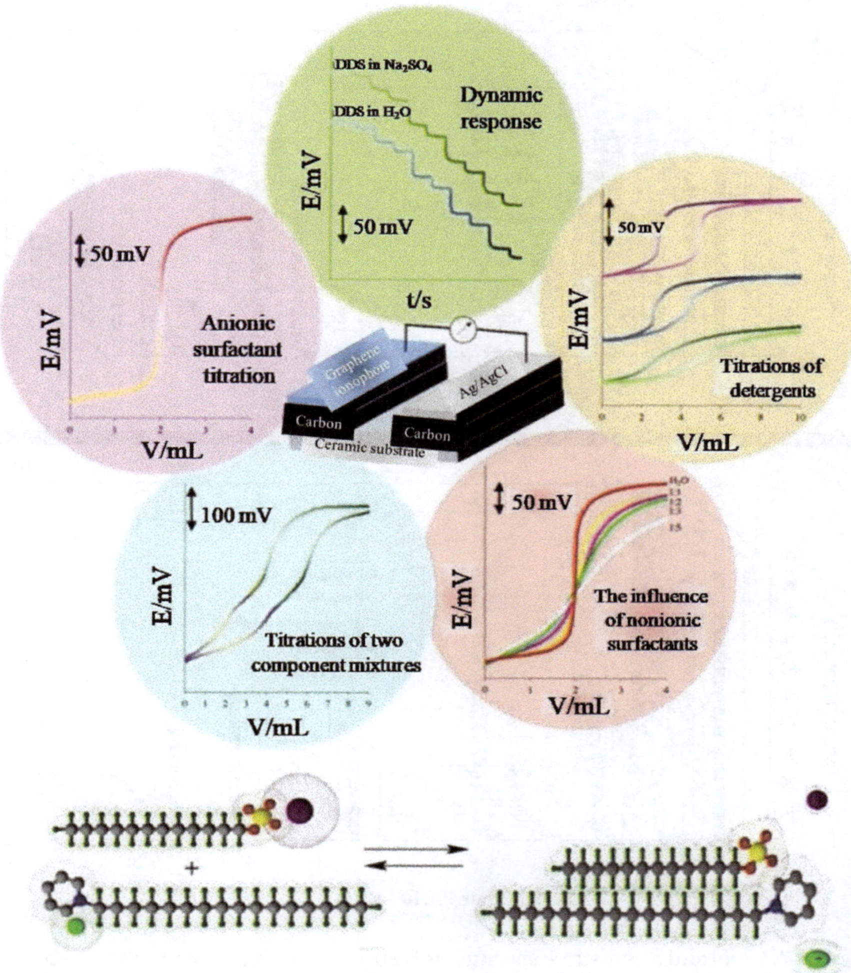

Figure 7.3 Summary of dynamic response and potentiometric titration response of an all solid-state sensor using dimethyldioctadecylammonium-tetraphenylborate (DDA-TPB). Reproduced from ref. 19 with permission from Elsevier, Copyright 2016.

7.3.1.3 Biosensors Based on Surfactants

In last few decades, different types of biosensors have been established to monitor or detect environmental pollutants. Some of the biosensors employ cells, microorganisms or enzymes to determine surfactant amount. A bioreactor-type electrochemical sensor was employed by Roger[25] to detect the presence of linear alkylbenzene sulphonates (LASs). The working principle behind this technology is the biodegradability of LASs by certain bacteria and alterations in their respiratory function in solution, associated with reduction in dissolved oxygen.

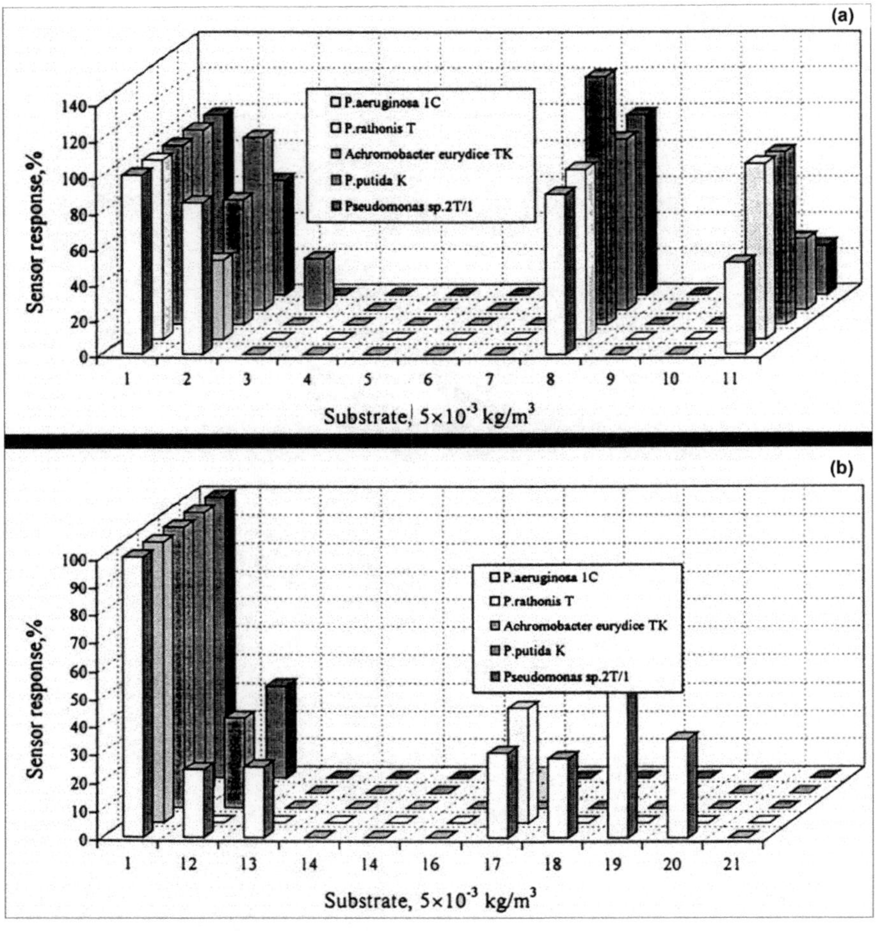

Figure 7.4 (a and b) Substrate specificity of sensors for surfactants: 1 – SDS; 2 – sodium alkansulphonate; 3 – sodium alkylmethyltaurine; 4 – alkylbenzenesulfonate (ABS1); 5 – ABS2; 6 – ABS3; 7 – ABS4; 8 – sulphosuccinic acid and disodium salt; 9 – alkylnapthlene sulphonate; 10 – decylbenzenesulphonate; 11 – alkoxymethyldiethylammonium methylsulphate; 12 – tween 80; 13 – tween 60; 14 – cetylpyridinium chloride; 15 – triton X-10; 16 – cetylether of polyethylene glycol ($n = 6$); 17 – dodecyclic ether of polyethylene glycol ($n = 10$); 18 – dodecyclic ether of polyethylene glycol ($n = 14$); 19 – tetradecyltrimethylammonium bromide; 20 – octoxinol 10 (OP 10); 21 – sulphobetain SB14. Reproduced from ref. 23 with permission from Elsevier, Copyright 2002.

This change in respiratory activity can be studied by implementation of an oxygen electrode, which is placed on-line along with an appropriate flow set-up. A linear response was obtained until 6 mg L^{-1} of LAS was degraded in less than 15 min. In another study, a *Pseudomonas rathonis* T-based amperometric biosensor was designed for detection of surfactants (anionic).

These microbes contained plasmid for the disintegration of the surfactant and exhibited greater sensitivity to DSD, *i.e.* within the range 0.25 to 0.75 mg L^{-1}. On the other hand, the response to other surfactants including volgonat, decylbenzene sulphonate, metaupon, toluene sulphonate and alkylbenzene sulphonate was 82%, 36%, 20%, 10% and 10%, respectively, as compared to SDS.[26] It has been observed that some of the surfactants inhibit cholinesterase (ChE) activity in aquatic animals and alter the chemical properties by modulating enzyme functionality through binding or rupturing the enzyme structure. In this regard, Garcia *et al.*[27] studied the effect of non-ionic surfactants on the activity of cholinesterase-based potentiometric biosensors. It is known that the response of any surface active compound results from the porosity and hydrophilic characteristic of enzyme support material and the inhibition mechanism. In their work, it was noticed that 0.002–0.3% (mass/volume ratio) of the surfactants displayed an inhibiting effect on the biosensor functionality. At low levels of concentration up to 10^{-4} mass/volume, the surfactants were capable of modifying the analytical features of irreversible and reversible inhibitor determination.

7.3.2 Surfactant Sensors for Environmental Analysis

Although some surfactants are regarded as pollutants, there are others which are eco-friendly and facilitate the efficient removal of other toxic agents in the environment. In the current scenario, harmful pollutants such as dyes, volatile organic compounds, heavy metals, pesticides and pharmaceuticals, waste generated from industry, personal care products *etc.*, pose danger to human health as well as other living organisms. Conventional methods employed to remove these pollutants are either cumbersome, less efficient or restricted owing to certain technoeconomic limitations. In this section, high pollutant removal efficacy of surfactants will be discussed.

7.3.2.1 Removal of Dyes

Nowadays, a number of synthetic dyes are used as food additives, and in cosmetics and drugs. In the food industry, these dyes tend to improve colour, flavour and desirability. In fact, synthetic dyes are preferred over natural dyes owing to less response to microbial contamination and pH, and minimal expenses during preparation. For instance, quinoline yellow (QY), a synthetic yellow colouring is widely used as a colourant in many drugs, food items, and cosmetic and hair products. However, QY is hazardous, causing skin allergies and irritation in children, respiratory problems, rashes and hyperactivity. There is also a probability of hypersensitivity in people having an allergy to aspirin. Knowing the above-mentioned conditions, the European Food Safety Committee has restricted the intake of QY from 10 to 0.5 mg kg^{-1} of body weight. In this regard, Mohseni-Shahri *et al.*[28] developed a simple and accurate detection technique for QY (Figure 7.5(a)) using a cationic surfactant, *i.e.* CTAB-decorated liquid crystal (LC)-based sensor.

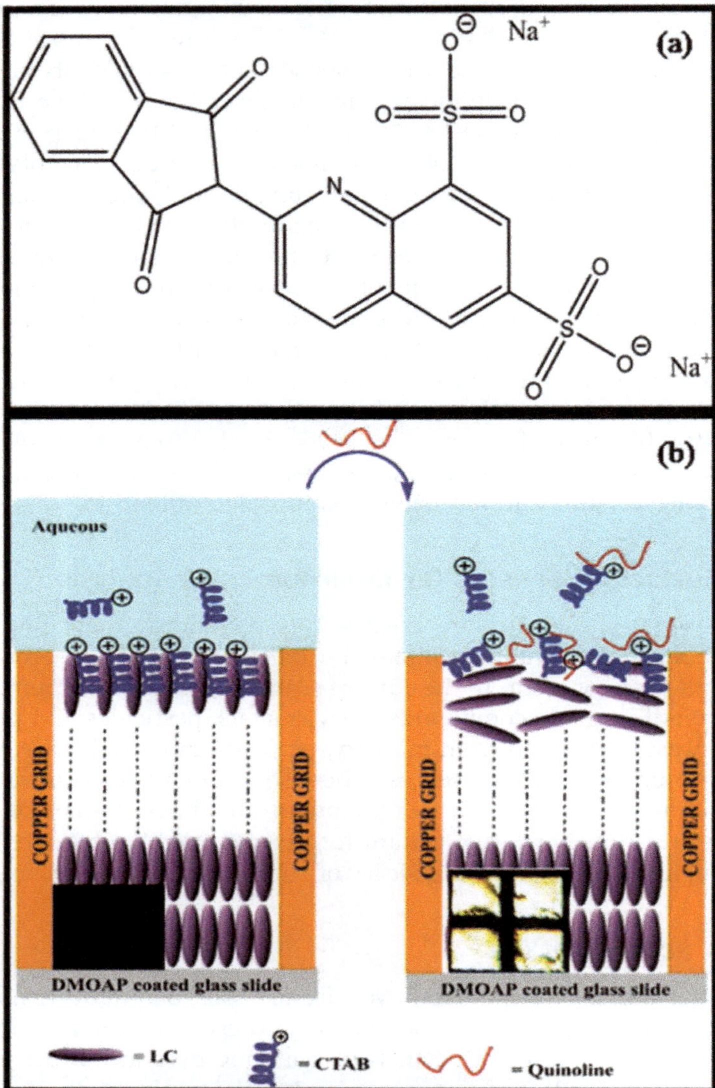

Figure 7.5 Schematic diagram showing (a) the basic structure of QY; (b) the strategy of the LC sensor to detect QY. Reproduced from ref. 28 with permission from Springer Nature, Copyright 2021.

The nematic LC in contact with CTAB revealed homeotropic alignment. On injecting quinoline yellow into the LC-cell, this homeotropic alignment was found to be altered to planar by electrostatic interactions between CTAB and QY. Such a design was useful in sensing QY at trace levels up to 0.5 fM with analogous selectivity. A schematic figure showing the strategy of the LC sensor for detection of QY is displayed in Figure 7.5(b). On the same note, surface modification of Se nanoparticles with a non-ionic surfactant, Brij-58 was employed for the selective adsorption of bromophenol blue (BB). These nanoparticles were not only biocompatible but also water soluble. Though nine different dyes were used as test materials for Brij-58-modified Se nanoparticles, BB exhibited maximum sensitivity and selectivity with adsorption up to 98% in only 6 minutes. The adsorption efficiency of BB was found to be 95%, even in real waste water samples.[29]

Dyes such as methylene blue (MB), crystal violet dye and rhodamine B (RhB) can also cause health hazards, even in small amounts. On skin contact, MB can lead to discolouration, a cold feeling and redness/dryness. Ingestion of methylene blue results in gastrointestinal irritation, discolouration of oral mucosa, irritation to the mouth, lips and throat, a pale complexion and a lack of coordination or drowsiness. Similarly, crystal violet dye belonging to the triphenyl carbocation group is widely employed for dying fabics like wool, cotton, silk and nylon. Being the brightest class of basic dye with tinctorial value it can cause human bladder cancer as well as cancer in the digestive system of other animals. In addition, RhB is also suspected to be carcinogenic. Removal of these dyes has been carried out using surface-modified commercial bentonite with hexadecyltrimethylammonium chloride surfactant (popularly known as organoclay). It was observed that higher pH led to higher adsorption for MB, crystal violet and RhB.[30]

7.3.2.2 Detection of Heavy Metals

The presence of heavy metals in the ecosystem poses a significant danger to living organisms, particularly water-dwelling species. These heavy metals, such as Cd^{2+}, Hg^{2+}, Sn^{2+}, Zn^{2+} and Sr^{2+}, *etc.*, are highly dangerous owing to their high affinity for ligands and chelating agents to form complexes. Such combinations lead to negative impacts on human health by altering the molecular structure of proteins and DNA and also act as inhibitors for different biocatalytic reactions. Amongst the top 12 toxic substances listed by the UN (United Nations), Cd and Cr are considered to be carcinogenic and the most lethal. While Hg is known to cause autism, vestibular dysfunction, birth defects, respiratory tract damage, kidney failure, skin diseases and speech impairment, Cd can lead to liver damage, modifications in blood and bone composition by destruction of red blood cells (RBCs), high blood pressure, replacement of calcium, skeletal disorders, *etc.* Working in this direction, Zahid *et al.*[31] developed a highly effectively electrochemical sensor using 1-cyclohexyl-3-dodecanoylthiourea (CDT) (surfactant) as the recognition layer. This electrochemical sensor was employed for the detection of Cd^{2+} and Hg^{2+} using chronocoulometry, impedance spectroscopy and also voltammetric techniques, including cyclic, square wave and differential pulse stripping voltammetry. The limits of detection of the modified electrode were lower than the danger level set by the World Health Organization (WHO),

Environmental Protection Agency (EPA) and European Water Quality (EWQ). The sensitivity of the electrode for Hg^{2+} and Cd^{2+} was found to be 0.034 μA L^{-1} $μg^{-1}$ and 0.010 μA L^{-1} $μg^{-1}$, respectively.

Previously, various other analytical techniques like inductively coupled plasma mass spectroscopy and UV-Vis spectrophotometry have been utilized for the detection of mercury. However, electrochemical methods modified with surfactants have been found to be more fast, reliable and cost effective, along with allowing detection down to trace levels. In this regard, glassy carbon electrodes are preferred owing to their anchoring ability and electrocatalytic activity. To further enhance their sensitivity, thiourea-based surfactant has been used for the surface modification of electrodes due to their unique interaction propensity for metal ions with a multitude of donor groups in their chemical structures. Again Zahid *et al.*[32] selected 1-(2,4-dinitrophenyl)-dodecanoylthiourea (DAN) with metal ions which acts as an electrocatalyst over a glassy carbon electrode. Under suitable experimental conditions (square wave voltammetry), a calibration plot showing good linearity up to 2 μg L^{-1} was obtained, with a detection limit of 0.64 μg L^{-1} and sensitivity of 0.164 μA L^{-1} $μg^{-1}$ in deionized (DI) water with an accumulation time of 360 seconds. Again in 2010, Sanchez and Walcarius[33] designed surfactant-templated silica films with immobilized 5-mercapto-1-methyl-tetrazole (MTTZ) groups (in a single step) coated over a glassy carbon electrode. The authors claimed that this thin-film electrode, constructed out of a homogeneous mesoporous layer functionalized with a sophisticated ligand and generated in a single step, was first of its kind. The film was highly porous and accumulated Hg(II) species very efficiently down to pH 1. The removal efficiency was greatly dependent on the functionalization level of the thin film and the optimum composition was noted to be 10% MTTZ. The detection efficiency with this model was found to be in the sub-micromolar concentration range down to a limit of 2×10^{-9} M (20 min accumulation). Recently, Shah and Shah[34] employed a more advanced way of detecting six different metal ions. They compared the electrochemical sensing platform using a 1-(3-chlorophenyl)-3-dodecanoylthiourea (Cl-DPTU)-modified glassy carbon electrode (GCE) with bare ones. Square wave anodic stripping voltammetry (SWASV) tests confirmed the sensing limits of Cl-DPTU/GCE under optimized conditions (6.45 nM, 7.85 nM, 9.15 nM, 11.0 nM and 178.8 nM for Cd^{2+}, Cu^{2+}, Hg^{2+}, Pb^{2+}, Zn^{2+} and Sr^{2+}, respectively) were better than bare GCEs. SWASV also confirmed the applicability of the electrodes for real water samples with 100% recovery. Figure 7.6 summarizes the repeatability, reproducibility and sensing ability of Cl-DPTU-modified glassy carbon electrodes.

7.3.2.3 Detection of Pesticides

Among the various types of pollutants, chemical pesticides used for agriculture tend to damage the ecosystem. A survey report says that the 1970s world pesticide market was governed by three categories of pesticides, namely, organophosphates, synthetic pyrethroids and carbamates. Pyrethroids and

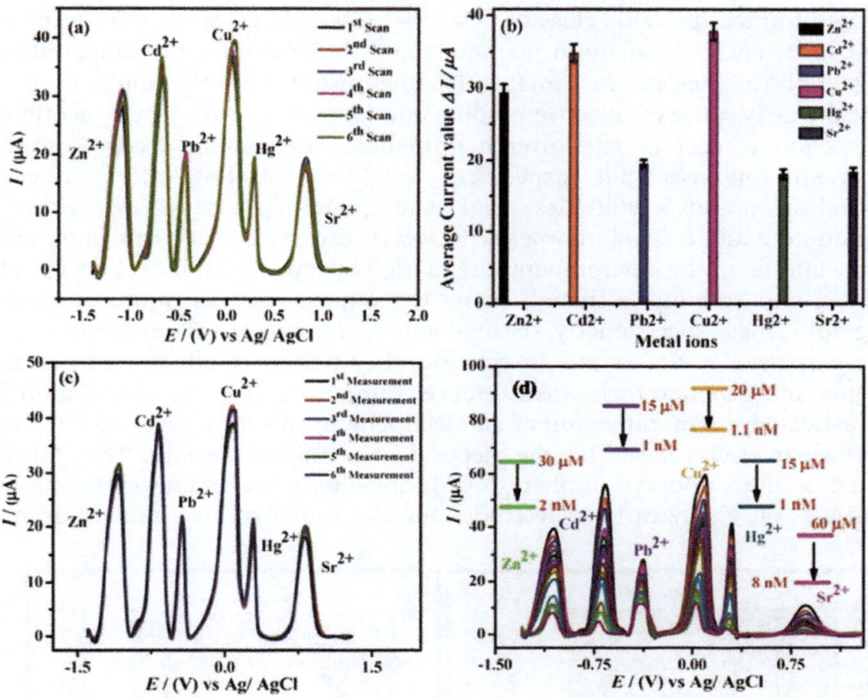

Figure 7.6 SWASV displaying (a) repeatability; (b) supporting bar graph; (c) reproducibility of Cl-DPTU modified glassy carbon electrodes and (d) sensing ability of different heavy metals at Cl DPTU/GCE in Britton–Robinson buffer (BRB) (scan rate ~100 mV s^{-1}, deposition potential ~−1.0 V, deposition time ~100 seconds; concentration range is presented above every peak). Reproduced from ref. 34 with permission from Elsevier, Copyright 2019.

organophosphate pesticides effectively control pest and diseases in different plant species.

But these pesticides are low in biodegradability and are highly persistent, due to which they are plentily available in air, water, sediments, soil, food and biological tissues. In addition, they are also known to be carcinogenic and can act as endocrine disrupters in several other organisms. In this regard, Sundar and Manisankar[35] synthesized a composite of multiwalled carbon nanotubes (MWCNTs) and PEDOT (poly(3,4-ethylenedioxythiophene) doped with surfactants (both anionic and cationic) to amalgamate the high sensing potential of PEDOT and antifouling ability of the surfactants. The constructed composite, when used with a glassy carbon electrode, demonstrated high sensitivity towards the reduction of cypermethrin (CYP), deltamethrin (DEL), fenvalerate (FEN), from the pyrethroid family, and dicofol (DCF) from the organochlorine family of pesticides. Besides, neonicotinoids, a category of neuro-active pesticides developed in 1980s are chemically related to nicotine. This class of pesticide possesses excellent effectiveness and relatively low toxicity to mammals and the environment compared to organophosphates and carbamates. Thiamethoxam, a neonicotinoid, is used worldwide because of selective binding and interaction with insect nicotinic acetycholine receptor sites over mammalian acetylcholine receptor sites. Therefore this insecticide is specifically used for control of sucking insects, including aphids, whiteflies, leaf and plant-hoppers, thrips, micro-lepidoptera and coleopteran pests. However, excessive usage can cause adverse effects on the environment due to high aqueous solubility (4.1 g L^{-1} at 25 °C) and stability. Although many techniques such as liquid chromatography, mass spectrometry, thermal lens spectrometry and immunoassays are employed to detect this insecticide, they require sophisticated instruments and generate toxic waste. Hence, Kumaravel and Chandrasekaran[36] investigated the modification of an electrochemical sensor electrode whose performance was based on the electrode–electrolyte interface. They fabricated sodium dodecyl sulphate (SDS), an anionic surfactant/silver nanoparticles on glassy carbon electrodes for electroanalysis of thiamethoxam.

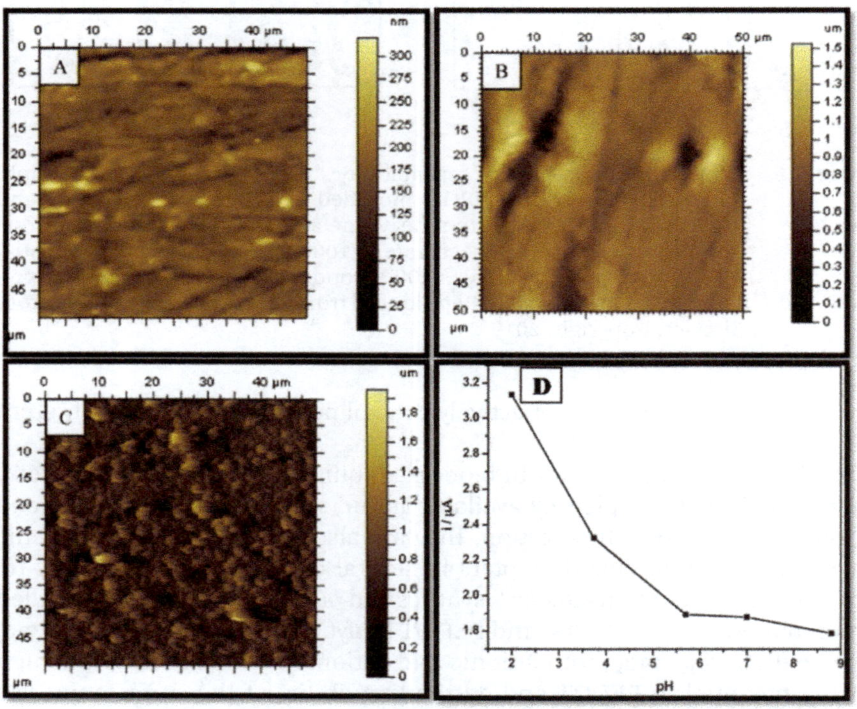

Figure 7.7 AFM images of (A) a bare GCE; (B) an SDS/GCE; (C) a silver nanoparticle/SDS/GCE; (D) peak current *vs.* pH for the reduction of thiamexotham on a silver nanoparticle/SDS/GCE electrode (sweep rate 100 mV per second). Reproduced from ref. 36 with permission from Elsevier, Copyright 2012.

Figure 7.7 compares AFM images of a bare GCE, an SDS/GCE and a silver nanoparticle/SDS/GCE. The electrodes displayed good electrochemical activity towards reduction of the pesticide, which was evident from the lower reduction potential and high sensing current. The electrode modification was quite simple and easily reproducible. Other than glassy carbon electrodes, polished silver solid amalgam electrodes (p-AgSAE) along with cetyltrimetyl ammonium bromide (a cationic surfactant) proved to be efficient to determine methomyl, a carbamate pesticide that is toxic in natural water samples. The cationic surfactant was found to regulate the kinetics and mechanism of methomyl reaction.[37]

7.4 Conclusion

The beneficial physicochemical properties of surfactants have been found to be suitable for worldwide industrial and household applications. In spite of their advantages, surfactants exhibit marked toxicity and are hazardous for human health and the environment. They can bind to proteins and modify the activity of various enzymes/other cell components, leading to malfunction. On the other hand, eco-friendly surfactants are utilized in many cases for the removal or detection of environmental pollutants such as colourants, heavy metals, insecticides and pesticides *etc.* Hence, in this chapter, efforts are made to focus on both of the above aspects of surfactants, just like two sides of a coin. In one section, surfactants are portrayed as pollutants and different sensing methods (particularly electrochemical sensors) for detecting these species down to trace level are presented. In the remaining section, eco-friendly surfactant-based sensors for tracing different environmental pollutants are presented. This chapter will definitely enrich the literature and will aid in the rational selection of surfactant sensors for environmental analysis.

References

1. S. H. Venhuis and M. Mehrvar, Health effects, environmental impacts, and photochemical degradation of selected surfactants in water, *Int. J. Photoenergy*, 2004, **6**(3), 115–125.
2. S. Rebello, A. K. Asok, S. Mundayoor and M. S. Jisha, Surfactants: chemistry, toxicity and remediation, in *Pollutant Diseases, Remediation and Recycling*, Springer, Cham, 2013, pp. 277–320.
3. G. G. Ying, Fate, behavior and effects of surfactants and their degradation products in the environment, *Environ. Int.*, 2006, **32**(3), 417–431.
4. M. Tandukar, S. Oh, U. Tezel, K. T. Konstantinidis and S. G. Pavlostathis, Long-term exposure to benzalkonium chloride disinfectants results in change of microbial community structure and increased antimicrobial resistance, *Environ. Sci. Technol.*, 2013, **47**(17), 9730–9738.
5. V. Abbot, V. Bhardwaj and P. Sharma, Investigation of intermolecular interactions of anionic surfactant SDS and rutin: A physico-chemical approach for pharmaceutical application, *J. Mol. Liq.*, 2021, **337**, 116352.

6. S. Ghosh and J. Dey, Interaction of bovine serum albumin with N-acyl amino acid based anionic surfactants: Effect of head-group hydrophobicity, *J. Colloid Interface Sci.*, 2015, **458**, 284–292.
7. R. Khalladi, O. Benhabiles, F. Bentahar and N. Moulai-Mostefa, Surfactant remediation of diesel fuel polluted soil, *J. Hazard. Mater.*, 2009, **164**(2–3), 1179–1184.
8. L. Wang, J. He, L. Zhu, Y. Wang, X. Feng, B. Chang, H. E. Karahan and Y. Chen, Assembly of pi-functionalized quaternary ammonium compounds with graphene hydrogel for efficient water disinfection, *J. Colloid Interface Sci.*, 2019, **535**, 149–158.
9. L. P. Singh, S. K. Bhattacharyya, G. Mishra and S. Ahalawat, Functional role of cationic surfactant to control the nano size of silica powder, *Appl. Nanosci.*, 2011, **1**(3), 117–122.
10. T. Förster, U. Issberner and H. Hensen, Lipid/surfactant compounds as a new tool to optimize skin-care properties of personal-cleansing products, *J. Surfactants Deterg.*, 2000, **3**(3), 345–352.
11. M. Zieba, D. Wieczorek, E. Klimaszewska, A. Malysa and D. Kwasniewska, Application of new synthesized zwitterionic surfactants as hair shampoo components, *J. Dispersion Sci. Technol.*, 2019, **40**(8), 1189–1196.
12. Q. Q. Zhang, B. X. Cai, W. J. Xu, H. Z. Gang, J. F. Liu, S. Z. Yang and B. Z. Mu, Novel zwitterionic surfactant derived from castor oil and its performance evaluation for oil recovery, *Colloids Surf., A*, 2015, **483**, 87–95.
13. C. Negin, S. Ali and Q. Xie, Most common surfactants employed in chemical enhanced oil recovery, *Petroleum*, 2017, **3**(2), 197–211.
14. M. He, K. Gao, L. Zhou, Z. Jiao, M. Wu, J. Cao, X. You, Z. Cai, Y. Su and Z. Jiang, Zwitterionic materials for antifouling membrane surface construction, *Acta Biomater.*, 2016, **40**, 142–152.
15. F. Bergaya and G. Lagaly, *General introduction: clays, clay minerals, and clay science*, Developments in Clay Science, 2006, vol. 1, pp. 1–18.
16. P. Li, A. Ghosh, R. F. Wagner, S. Krill, Y. M. Joshi and A. T. Serajuddin, Effect of combined use of nonionic surfactant on formation of oil-in-water microemulsions, *Int. J. Pharm.*, 2005, **288**(1), 27–34.
17. P. U. Sap, D. F. Anghel and C. Luca, Electrochemical sensors for anionic surfactants. 1. The electrode based on bis[2,4,6-tri(2-pyridyl)-1,3,5-triazine] iron (ii) dodecyl-sulfate, *Rev. Roum. Chim.*, 1983, **28**(9–10), 883–890.
18. M. Arvand-Barmchi, M. F. Mousavi, M. A. Zanjanchi and M. Shamsipur, A new dodecylsulfate-selective supported liquid membrane electrode based on its N-cetylpyridinium ion-pair, *Microchem. J.*, 2003, **74**(2), 149–156.
19. O. Galović, M. Samardžić, M. Hajduković and M. Sak-Bosnar, A new graphene-based surfactant sensor for the determination of anionic surfactants in real samples, *Sens. Actuators, B*, 2016, **236**, 257–267.
20. M. Skoog, K. Kronkvist and G. Johansson, Blocking of chemically modified graphite electrodes by surfactants, *Anal. Chim. Acta*, 1992, **269**(1), 59–64.

21. S. B. Adeloju and S. J. Shaw, Indirect determination of surfactants by adsorptive voltammetry: Part 1: determination of trace and ultratrace concentrations of sodium dodecylbenzene sulfonate, *Electroanalysis*, 1994, **6**(8), 639–644.
22. S. B. Adeloju and S. J. Shaw, Indirect determination of surfactants by adsorptive voltammetry: Part 2: Determination of hexadecyl trimethyl ammonium bromide and cetyl pyridinium chloride in industrial and consumer products, *Electroanalysis*, 1994, **6**(8), 645–649.
23. L. Taranova, I. Semenchuk, T. Manolov, P. Iliasov and A. Reshetilov, Bacteria-degraders as the base of an amperometric biosensor for detection of anionic surfactants, *Biosens. Bioelectron.*, 2002, **17**(8), 635–640.
24. E. Khaled and H. Y. Aboul-Enein, Surfactants, in *Environmental Analysis by Electrochemical Sensors and Biosensors: Applications*, 2015, pp. 905–930.
25. K. R. Rogers, Recent advances in biosensor techniques for environmental monitoring, *Anal. Chim. Acta*, 2006, **568**(1–2), 222–231.
26. Y. Nomura, K. Ikebukuro, K. Yokoyama, T. Takeuchi, Y. Arikawa, S. Ohno and I. Karube, A novel microbial sensor for anionic surfactant determination, *Anal. Lett.*, 1994, **27**(15), 3095–3108.
27. L. M. Garcia, B. Castro, R. Ribeiro and L. Guilhermino, Characterization of cholinesterase from guppy (Poecilia reticulata) muscle and its in vitro inhibition by environmental contaminants, *Biomarkers*, 2000, **5**(4), 274–284.
28. F. S. Mohseni-Shahri, F. Moeinpour and A. Verdian, A cationic surfactant-decorated liquid crystal-based sensor for sensitive detection of quinoline yellow, *Sci. Rep.*, 2021, **11**(1), 24264.
29. P. Chauhan, K. K. Bhasin and S. Chaudhary, High selectivity and adsorption proficiency of surfactant-coated selenium nanoparticles for dye removal application, *Environ. Sci. Pollut. Res.*, 2021, **28**(43), 61344–61359.
30. T. S. Anirudhan and M. Ramachandran, Adsorptive removal of basic dyes from aqueous solutions by surfactant modified bentonite clay (organoclay): kinetic and competitive adsorption isotherm, *Process Saf. Environ. Prot.*, 2015, **95**, 215–225.
31. A. Zahid, A. Shah, F. J. Iftikhar, A. H. Shah and R. Qureshi, Surfactant modified glassy carbon electrode as an efficient sensing platform for the detection of Cd(II) and Hg(II), *Electrochim. Acta*, 2017, **235**, 72–78.
32. A. Zahid, A. Lashin, U. A. Rana, N. Al-Arifi, I. Ullah, D. D. Dionysiou, R. Qureshi, A. Waseem, H. B. Kraatz and A. Shah, Development of surfactant based electrochemical sensor for the trace level detection of mercury, *Electrochim. Acta*, 2016, **190**, 1007–1014.
33. A. Sanchez and A. Walcarius, Surfactant-templated sol–gel silica thin films bearing 5-mercapto-1-methyl-tetrazole on carbon electrode for Hg(II) detection, *Electrochim. Acta*, 2010, **55**(13), 4201–4207.
34. A. Munir, A. Shah, J. Nisar, M. N. Ashiq, M. S. Akhter and A. H. Shah, Selective and simultaneous detection of Zn^{2+}, Cd^{2+}, Pb^{2+}, Cu^{2+}, Hg^{2+}

and Sr^{2+} using surfactant modified electrochemical sensors, *Electrochim. Acta*, 2019, **323**, 134592.
35. P. L. Abirama Sundari and P. Manisankar, Development of ultrasensitive surfactants doped poly(3, 4-ethylenedioxythiophene)/multiwalled carbon nanotube sensor for the detection of pyrethroids and an organochlorine pesticide, *J. Appl. Electrochem.*, 2011, **41**, 29–37.
36. A. Kumaravel and M. Chandrasekaran, Nanosilver/surfactant modified glassy carbon electrode for the sensing of thiamethoxam, *Sens. Actuators, B*, 2012, **174**, 380–388.
37. C. C. G. Silva and D. D. Souza, Polished silver solid amalgam electrode and cationic surfactant as tool in electroanalytical determination of methomyl pesticide, *Talanta*, 2018, **189**, 389–396.

CHAPTER 8

Surfactant Sensors for Food Quality Monitoring

SANDYA TAMBI DORAI, PRIYA TIWARI, YASHI DWIVEDI AND SANDEEP CHANDRASHEKHARAPPA*

Department of Medicinal Chemistry, National Institute of Pharmaceutical Education and Research (NIPER-R), Raebareli, Transit Campus, Bijnor-Sisendi Road, Sarojini Nagar, Near CRPF Base Camp, Lucknow (UP)-226002, India
*Emails: c.sandeep@niperraebareli.edu.in; c.sandeep@niperrbl.ac.in

8.1 Introduction

The process of reviewing and assuring the safety and quality of food items along the production, distribution, and consumption chain is known as food quality monitoring.[1] It entails testing and analyzing food samples to verify that dangerous contaminants such as bacteria, viruses, and toxins are not present and fulfill nutritional, taste, texture, and aesthetic criteria.[1–3] Food quality monitoring is critical for public health, avoiding foodborne diseases, and ensuring that customers have access to safe and high-quality food items. Monitoring of food has become more sophisticated, precise, and efficient in recent years as technology has advanced and customer demand for clarity in the food supply chain has increased.[4,5]

8.2 Surfactants

Surfactants are often called surface-active agents, which are chemicals that play an important role in a wide range of commercial and home goods, including soaps, detergents, shampoos, and cleaning agents.[6] They are

Detection Science Series No. 23
Surfactant-based Sensors in Chemical and Biochemical Detection
Edited by J. G. Manjunatha

Published by the Royal Society of Chemistry, www.rsc.org

amphiphilic molecules, which means they contain both hydrophilic and hydrophobic characteristics. Surfactants reduce the surface tension between two liquids or between a liquid and a solid, which allows them to readily mix and remove dirt, oil, and other impurities, as they readily produce micelles in aqueous solutions. Micelles are spherical clusters of surfactant molecules with the hydrophobic tail pointing inwards and hydrophilic heads pointing towards the outside facing the aqueous media.[4,7]

8.2.1 Classification of Surfactants

i. Anionic (negatively charged hydrophilic groups)
ii. Cationic (positively charged hydrophilic groups)
iii. Amphoteric (both positively and negatively charged hydrophilic groups)
iv. Non-ionic based on the charge on the hydrophilic group (no charge on hydrophilic groups).[6]

8.3 Sensors

The detection and measurement of numerous characteristics relating to food quality, safety, and freshness are common tasks for sensors in food quality monitoring.[8] These sensors may be included in various food processing and packaging systems and can be categorized according to the type of measurements carried out.[9–11]

8.3.1 Chemical Sensors

These types of sensors are commonly used in food quality monitoring to detect specific chemical compounds, such as gases or volatile organic compounds (VOCs), that can be indicative of food spoilage or contamination.[12]

8.3.2 Physical Sensors

These sensors measure physical properties such as temperature, pressure, or moisture content, which can affect food quality and safety.[8]

8.3.3 Biological Sensors

Biological sensors are mainly used in food quality monitoring, which detects the presence of microorganisms, such as bacteria or viruses that can cause foodborne illness or spoilage.[12]

8.3.4 Next Generation Sensors

The next generation of technology can help to track the state of the goods with smart sensors and labels that can be affixed to the packaging.[13] They

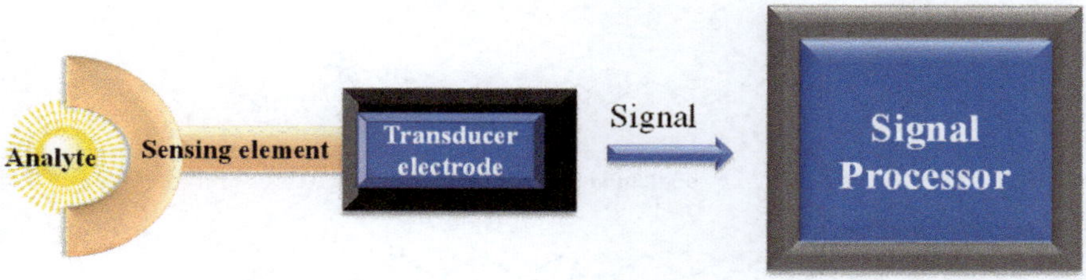

Figure 8.1 Flow of events in an electrochemical sensor.

can be created to track freshness indicators and give a real-time "index of quality" for the product, track temperature fluctuations, or spot potentially dangerous substances. Additional abilities can be added to provide protective activities, such as a coating on the packaging that can operate as an oxygen barrier to prevent rotting.[14] For a long period of time, sensors were made up of chemical or biological receptors that were created particularly to detect a target analyte and a physical transducer that transformed the recognition into a signal (Figure 8.1 shows the electrochemical sensing of an analyte) that could be measured and produced a qualitative/quantitative output.[15]

While designing efficient food sensors four classes of analytes are usually used.

i. Chemical and biological contaminants[16]
ii. Allergens[17]
iii. Nutritional constituents[18]
iv. Food additives that when used in excess could be detrimental.[19]

The freshness of a desirable food product can be assessed by the identification of specific analytes. Food quality and safety sensors may be built into wearable devices, miniaturized devices, or even the food packaging itself. Examples of recently created food sensor are a glove with sensors that can be used as a wearable device to detect food flavor and the incorporation of sensors into a portable device for the detection of food allergens.[20,21]

8.4 Role of Surfactants in Monitoring Food Quality

Surfactants are a class of amphiphilic ions or molecules known as surfactants that have a lengthy hydrophobic tail on one side and a hydrophilic head on the other. These, when adhered to the electrode surface, change the characteristics of the electrode–solution interface and subsequently influence electrochemical reactions involving electroactive species. Surfactants are frequently employed in chemistry and have a significant impact on various electrochemical processes. The outcomes have demonstrated that some surfactants significantly boost electrochemical responses to the analyte.[22] Moreover, the disposable electrode has been frequently altered with

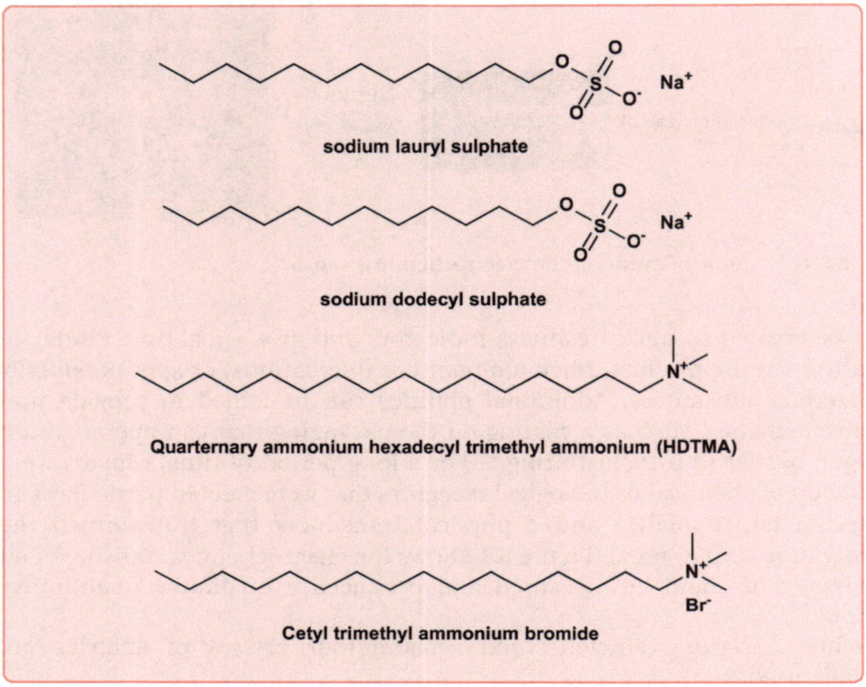

Figure 8.2 Commonly used surfactants.

various materials to determine a variety of analytes (Figure 8.2 shows the most-used surfactants):

i. Sodium lauryl sulphate
ii. Sodium dodecyl sulphate
iii. Quarternary ammonium hexadecyl trimethyl ammonium (HDTMA)
iv. Cetyl trimethyl ammonium bromide (CTAB).[22]

Surfactants are also significant in food quality control and are very commonly employed in the food industry to aid the extraction and analysis of food samples. They also enhance the sensitivity and selectivity of analytical methods used in quality monitoring. For instance, surfactants may be added to food samples prior to analysis to improve the accuracy and precision of tests for the presence of contaminants such as pesticides and heavy metals.[23] The use of surfactants in antimicrobial treatments increases the efficacy of preservatives and prevents the growth of harmful microorganisms in food products.[24]

8.4.1 Advantages of Surfactant-modified Sensors

i. Increased sensitivity
ii. Enhancement of potential to amplify signals[25]
iii. Higher transfer rate of electrons[26]

iv. Amendment of redox potential and increase of peak current leads to a change in stability of electro-generated intermediates and electrochemical products.
v. Anti-fouling problems are surmounted.[27]

8.5 Application of Surfactants in Food Quality Monitoring

8.5.1 Surfactants Used in the Detection of Allergens in Food Products

Food allergen control is a critical issue since numerous individuals globally, *i.e.*, 1–2% of adults and 5–8% of children, suffer from food allergies. To maintain food safety, food manufacturers must indicate allergy information on their goods, stating the presence of food allergens in food items. According to US and Japan regulations it is mandatory to include seven different categories of food. In Japan if the food contains more than 10 gg^{-1} of protein, labeling of such food products is required. Hideshimaa *et al.* studied the detection of allergenic buckwheat protein (BWp16) using a label-free field-effect transistor (FET) biosensor with the use of anionic surfactants. Proteins which are neutral in nature cannot be detected by the sensors, therefore to increase the sensitivity of the detection, surfactants are useful. In order to achieve detection, they employed different concentrations of surfactants, SDS solution (0.01, 0.1, 0.5, 1, 2% w/w), CTAB (1% w/w), and ODG (n-octyl-β-d-glucoside) (1% w/w). Surfactants increase the net charge on allergens which lack sufficient inherent charge to be picked up by the sensor, therefore enhancing the signal by conjugating the charged surfactant with the protein. Among all the surfactants tested SDS (1% w/w) increased the detection of the buckwheat protein in the FET biosensor.

8.5.2 Detection of Vanillin in Food, Beverages, and Pharmaceuticals

Vanillin (4-hydroxy-3-methoxybenzaldehyde) is the main chemical component of vanilla bean extract. It is the most commonly used flavoring agent in many foods and beverages. About 99% of vanillin (VAN) is synthetically produced and 1% is obtained from natural sources like guaiacol, petrochemical raw components, wood, and biomass (organic materials from plants), and it has been widely extracted from a lignin source.[28,29]

VAN contains a wide range of applications:

- Antioxidant properties are majorly used in the preservation of milk and its by-products.
- Treatment of various disorders like cancer, sickle cell anemia, and cardiac disease.

Inspite of its vast application it possesses some side effects, such as headache, nausea, damage to the liver and kidney, dyspnea, and leads to

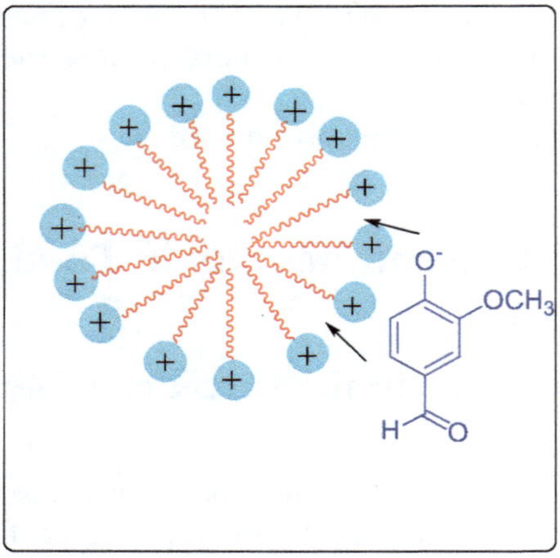

Figure 8.3 Interaction of positively charged cationic CTAB with negatively charged VAN.

addiction to food or beverages containing vanillin.[30] Hence detection of vanillin content in food, beverages, and pharmaceutical products is important. Veera Erady *et al.* developed a modified carbon paste electrode cast using cetyltrimethylammonium bromide (CTAB) with differential pulse voltammetry techniques for the detection of vanillin in food products. The surfactant CTAB provides a favorable environment for detection because of the spherical rod-like micelles it forms and the hydrophobic interaction with negatively charged VAN leads to a strong C–H chain on the electrode surface which forms a compact monolayer where the polar heads are towards the aqueous phase (pH 7.38). Figure 8.3 shows the interaction of positively charged cationic CTAB with negatively charged VAN. As a result, CTAB increases the current and reduces the overpotential, which leads to sensing of VAN by the detector.[31]

8.5.3 Detection of Adulterants by Surfactant-coated Nanoparticles

Surfactant-coated nanoparticles are used in food for sensing artificial smell, taste, bacterial toxins, freshness, and pathogenic bacteria. This technique plays a significant role in improving the evaluation of food quality and safety. Table 8.1 indicates the list of surfactants used and their applications.[32]

8.5.4 Surfactants Used in Packaging of Food Products

Surfactant-modified cellulose nanocrystal embedded in the thermoplastic polymer polylactic acid (PLA), generated from natural sources was used to create nano-biocomposites. The modified cellulose nanocrystal had an

Table 8.1 Nanoparticle-coated surfactants.

Surfactant	Core	Application	Ref.
Sodium dodecyl sulphate or polysorbate 80	Silver	Antimcrobial activity	Kvitek *et al.*[33]
Cationic surfactant	Silica-modified magnetite	Detection of limits of Cu(II), Ni(II), Co(II), Cd(II), Pb(II) and Mn(II) in aqueous solution	Karatapanis *et al.*[34]
(1-(2,4-Dinitrophenyl)-dodecanoylthiourea (DAN)	Electrochemical sensor	Detection of Hg(II) in drinking water	Zahid *et al.*[35]
Poloxamer 407	Cocoa butter	Enhanced oxidative and thermal stability of linoleic acid in water	Mengqian Gao *et al.*[36]
Sophorolipid	Zein	Decreased degradation of lutein in water	Yongkai Yuan *et al.*[37]

appropriate dispersion medium in the surfactant, which increases the homogeneity of the film, thereby resulting in better oxygen barrier characteristics and a 34% decrease in water permeability.[38,39]

8.5.5 Surfactants Used in Voltammetric Sensors

Voltammetric sensors measure current by applying a changing potential to the sensor. This type of sensor typically uses a three-electrode system, including a working electrode, a reference electrode, and an auxiliary electrode. The flow of electrons caused by the oxidation or reduction of the electroactive species is what creates the current. As a result, the current is inversely proportional to the analyte concentration in the sample. There are several techniques to change the potential profile, including linear sweep, cyclic, and square wave.[40] Amperometric sensors can be thought of as a subset of voltammetric sensors because they produce the same output signal but maintain a constant potential using an external energy source, such as batteries. As an illustration, a voltammetric food sensor was created using modified carbon paste electrodes to find samples of dried fruit and soft drinks that included the food colors carmoisine and tartrazine, which can be toxic to children. With low detection limits of 20 nM for carmoisine and 0.06 μM for tartrazine, these voltammetric sensors were able to concurrently identify carmoisine and tartrazine in the ranges 70–650 μM and 0.1–750 μM, respectively.[41]

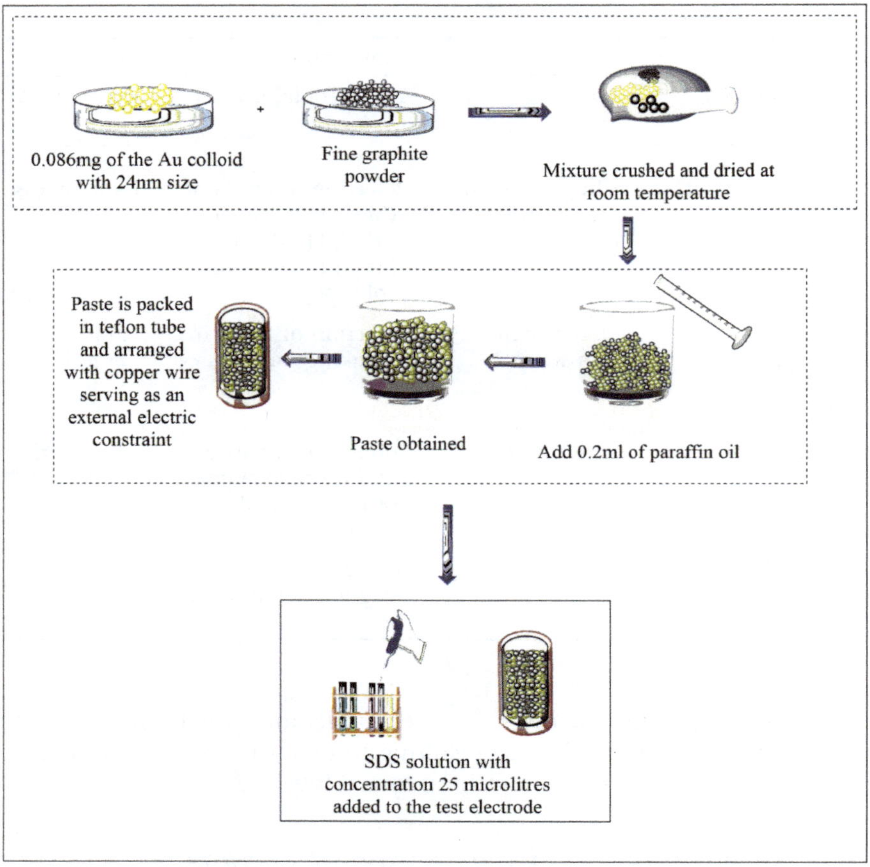

Figure 8.4 General method for the preparation of the graphite-based gold electrode and surfactant used in electrosensing studies.

8.5.5.1 Detection of Tyrosine in Food Products

Tyrosine (Tyr), which is abundant in foods like soymeals and cereals, was measured using a highly sensitive sensor made from a carbon paste electrode modified with gold nanoparticles (AuMCPE) in the presence of the anionic surfactant sodium dodecyl sulphate (SDS) in an aqueous solution. Preparation of the electrode and the methodology of detection *via* the SDS-modified electrode is described in Figure 8.4. Differential pulse voltammetry (DPV), cyclic voltammetry (CV), chronocoulometry, and chronoamperometry techniques were used to conduct the measurements. In the presence of SDS, the produced electrode exhibited voltammetric responses that were very sensitive and selective for Tyr. The developed sensor was found efficient with satisfactory results for the determination of Tyr, specifically in the presence of sodium dodecyl sulphate (SDS) as an anionic surfactant. The simple and

Table 8.2 Electrochemical sensors based on the use of surfactants.

Technique	Surfactant	Working electrode	Analyte	Practice sample	LOD	Linear range	Ref.
CV	SDS	SDS/GPE	CC	Tap water	0.10 µM	2–10 µM 15–150 µM	Manjunatha (2020)[43]
CV	TX-100	TX-100/CPE	IG	Chocolate local water	0.16 µM	1.5–2 µM	Raril *et al.* (2018)[44]
SWV	CTAB	HMDE	Genistein	Soy flour	34.3 nM	0.11–1.09 µM	Fogliatto *et al.* (2010)[45]
SWV	SDS	SDS/CPE	Ethylvanillin Methylvanillin	Foodstuffs	0.02 µM Ethylvanillin 0.02 µM Methylvanillin	1–20 µM Ethylvanillin 0.07–0.2 µM Methylvanillin	Farag *et al.* (2018)[46]
SWAdSV	CTAB	CPT-BDD	Catechol-containing quercetin	Apple juice	0.44 nM	1.7–0.33 µM	Abdullah *et al.* (2018)[47]

economic method of preparation of these electrode sensors with satisfactory results suggests that the developed electrodes are an attractive candidate for food monitoring and for other practical assessments.[42] There are different surfactants used as electrochemical sensors for food monitoring and some of them are discussed below in Table 8.2.

8.5.6 Detection of Quinoline Yellow (QY)

Quinoline yellow is a synthetic yellow color dye used in food products, drugs, soft drinks, and many cosmetic products. It has been seen in various studies that Quinoline yellow causes skin allergies, irritation, respiratory symptoms, hyperactivity, and skin rashes.[48] Hence, its detection is important for reducing risk. In the reported literature, a liquid crystal (LC)-based sensor was developed for sensitive detection of QY with the assistance of a cationic surfactant. It was found that the liquid crystal showed homeotropic alignment with CTAB, but when QY was injected into the LC cell, the alignment changed to a planar form due to electrostatic interactions between QY and CTAB. It can detect even at a trace level of about 0.5 fM with analogous selectivity. This cationic surfactant-decorated LC-based sensor is a sensitive, rapid, convenient, and simple piece of equipment for label-free detection of QY in the food industry for safety measures.[48]

8.5.7 Curcumin Detection in Natural Food Supplements Through a Surfactant-modified Voltammetric Sensor

Curcumin is a substance obtained from the rhizome of *Curcuma longa* L. that is widely used as a food additive and in other dyeing applications. It is known for its antioxidant, anti-inflammatory, anticarcinogenic, and anti-HIV properties. Likewise, it is also used in the treatment of Alzheimer disease, multiple sclerosis, and dementia.[49] Many analytical methods such as UV-spectrophotometry, HPLC, and capillary electrophoresis are involved in the isolation of curcumin, but due to their simplicity, sensitivity, and low-cost, electrochemical techniques have proved better alternatives. Furthermore, it has been observed that surfactant-modified electrodes increase the efficiency of the technique. In the literature, sodium dodecyl sulphate-modified carbon nanocomposite material was used to detect curcumin in the presence of vitamin B_2. Sodium dodecyl sulphate acts as a redox mediator. This modification of an carbon nanocomposite paste electrode provides high sensitivity and low detection limits. The developed sensor has a low detection limit equal to 2.7×10^{-8} M for curcumin and shows good linearity.[49]

8.6 Conclusion

Assessing the quality of food using surfactants is crucial. In the food industry they are widely used due to properties such as emulsification, foaming, and wetting. The application of surfactants in food quality monitoring is critical for safeguarding the security and quality of the food supply. They assist in extracting and solubilizing pollutants from food metrics, which makes it simpler to find them in analytical procedures. Tests for the safety of food also use surfactants, which increase the sensitivity of microbial detection techniques in identifying pathogenic microorganisms. Overall, surfactants help in determining the quality of food by aiding in the analysis and preservation of food products, as well as in improving the accuracy and sensitivity of analytical methods to ensure food safety and quality.

References

1. A. Prajwal, P. Vaishali, Z. Payal and D. Sumit, Food Quality Detection and Monitoring System, 2020 IEEE International Students' Conference on Electrical, Electronics and Computer Science, SCEECS 2020, Institute of Electrical and Electronics Engineers Inc., 2020.
2. G. S. Kumari, Food Quality Monitoring System by using Arduino. Available from: www.jespublication.com/.
3. S. Abasi, S. Minaei, B. Jamshidi and D. Fathi, Dedicated non-destructive devices for food quality measurement: A review, *Trends Food Sci. Technol.*, 2018, **78**, 197–205.
4. J. Jeevanandam, D. Agyei, M. K. Danquah and C. Udenigwe, Food quality monitoring through bioinformatics and big data, *Future Foods: Global Trends, Opportunities, and Sustainability Challenges*, 2022, pp. 733–744.
5. R. N. Pudake, N. Chauhan and C. Kole, *Nanoscience for Sustainable Agriculture*, 2019.
6. R. Sharma, Surfactants: Basics and Versatility in Food Industries. Available from: https://www.researchgate.net/publication/328065896.
7. T. Miyazawa, M. Itaya, G. C. Burdeos, K. Nakagawa and T. Miyazawa, A Critical Review of the Use of Surfactant-Coated Nanoparticles in Nanomedicine and Food Nanotechnology, *Int. J. Nanomed.*, 2021, **16**, 3937–3999.
8. F. Oveissi, L. H. Nguyen, J. E. Giaretta, Z. Shahrbabaki, R. J. Rath and V. A. Apalangya, *et al.*, Sensors for food quality and safety, *Food Engineering Innovations Across the Food Supply Chain*, 2022, pp. 389–410.
9. F. S. Mohseni-Shahri, F. Moeinpour and A. Verdian, A cationic surfactant-decorated liquid crystal-based sensor for sensitive detection of quinoline yellow, *Sci. Rep.*, 2021, **11**(1), 1–7. Available from: https://www.nature.com/articles/s41598-021-03788-6.
10. M. M. Charithra, J. G. G. Manjunatha and C. Raril, Surfactant Modified Graphite Paste Electrode as an Electrochemical Sensor for the Enhanced Voltammetric Detection of Estriol with Dopamine and Uric acid, *Adv. Pharm. Bull.*, 2020, **10**(2), 247–253. Available from: DOI: 10.34172/apb.2020.029.
11. F. Garkani Nejad, S. Tajik, H. Beitollahi and I. Sheikhshoaie, Magnetic nanomaterials based electrochemical (bio)sensors for food analysis, *Talanta*, 2021, **228**, 122075.

12. F. Mustafa and S. Andreescu, Chemical and Biological Sensors for Food-Quality Monitoring and Smart Packaging, *Foods*, 2018, **7**(10), 168, DOI: 10.3390/foods7100168.
13. R. Dobrucka, Metal nanoparticles in nanosensors for food quality assurance, *LogForum*, 2020, **16**(2), 271–278.
14. F. Mustafa and S. Andreescu, Chemical and Biological Sensors for Food-Quality Monitoring and Smart Packaging, *Foods*, 2018, **7**(10), 168. Available from: https://www.mdpi.com/2304-8158/7/10/168/htm.
15. Smart packaging improving shelf life of food product. Available from: https://www.researchgate.net/publication/337338435.
16. G. A. Salvatore, J. Sülzle, F. Dalla Valle, G. Cantarella, F. Robotti and P. Jokic, *et al.*, Biodegradable and Highly Deformable Temperature Sensors for the Internet of Things, *Adv. Funct. Mater.*, 2017, **27**(35), 1702390. Available from: https://onlinelibrary.wiley.com/doi/full/10.1002/adfm.201702390.
17. M. Zhao, Y. Li, X. Ma, M. Xia and Y. Zhang, Adsorption of cholesterol oxidase and entrapment of horseradish peroxidase in metal-organic frameworks for the colorimetric biosensing of cholesterol, *Talanta*, 2019, **200**, 293–299.
18. Y. J. Thong, T. Nguyen, Q. Zhang, M. Karunanithi and L. Yu, Predicting food nutrition facts using pocket-size near-infrared sensor, Proceedings of the Annual International Conference of the IEEE Engineering in Medicine and Biology Society, EMBS, 2017, pp. 742–745.
19. J. G. Manjunatha, A novel voltammetric method for the enhanced detection of the food additive tartrazine using an electrochemical sensor, *Heliyon*, 2018, **4**(11), e00986.
20. R. K. Mishra, L. J. Hubble, A. Martín, R. Kumar, A. Barfidokht and J. Kim, *et al.*, Wearable Flexible and Stretchable Glove Biosensor for On-Site Detection of Organophosphorus Chemical Threats, *ACS Sens.*, 2017, **2**(4), 553–561. Available from: https://pubmed.ncbi.nlm.nih.gov/28723187/.
21. B. Ciui, A. Martin, R. K. Mishra, T. Nakagawa, T. J. Dawkins and M. Lyu, *et al.*, Chemical Sensing at the Robot Fingertips: Toward Automated Taste Discrimination in Food Samples, *ACS Sens.*, 2018, **3**(11), 2375–2384. Available from: https://pubmed.ncbi.nlm.nih.gov/30226368/.
22. S. M. Ghoreishi, M. Behpour, N. Jafari and A. Khoobi, Determination of Tyrosine in the Presence of Sodium Dodecyl Sulfate Using a Gold Nanoparticle Modified Carbon Paste Electrode, *Anal. Lett.*, 2013, **46**(2), 299–311. Available from: https://www.tandfonline.com/doi/abs/10.1080/00032719.2012.710867.
23. F. Mustafa and S. Andreescu, Chemical and biological sensors for food-quality monitoring and smart packaging, *Foods*, 2018, **7**(10), 168.
24. L. Rodrigues, I. M. Banat, J. Teixeira and R. Oliveira, Biosurfactants: Potential applications in medicine, *J. Antimicrob. Chemother.*, 2006, **57**, 609–618.
25. J. S. Farias, H. Zanin, A. S. Caldas, C. C. dos Santos, F. S. Damos and R. de Cássia Silva Luz, Functionalized Multiwalled Carbon Nanotube Electrochemical Sensor for Determination of Anticancer Drug Flutamide, *J. Electron. Mater.*, 2017, **46**(10), 5619–5628.

26. P. K. Brahman, N. Pandey, J. V. S. Kumar, P. Somarouthu, S. Tiwari and K. S. Pitre, Highly sensitive stripping voltammetric determination of a biomolecule, pyruvic acid in solubilized system and biological fluids, *Arabian J. Chem.*, 2016, **9**, S1897–S1904.
27. L. Zhu, Y. Cao and G. Cao, Electrochemical sensor based on magnetic molecularly imprinted nanoparticles at surfactant modified magnetic electrode for determination of bisphenol A, *Biosens. Bioelectron.*, 2014, **54**, 258–261.
28. I. Labuda, Flavor Compounds, *Encyclopedia of Microbiology*, 2009, pp. 305–320.
29. M. Fache, B. Boutevin and S. Caillol, Vanillin Production from Lignin and Its Use as a Renewable Chemical, *ACS Sustainable Chem. Eng.*, 2016, **4**(1), 35–46. Available from: https://pubs.acs.org/doi/full/10.1021/acssuschemeng.5b01344.
30. Ethyl vanillin – toxicity, side effects, diseases, and environmental impacts. Available from: https://naturalpedia.com/ethyl-vanillin-toxicity-side-effects-diseases-and-environmental-impacts.html.
31. V. Erady, R. J. Mascarenhas and A. K. Satpati, Highly efficient and selective quantification of vanillin in food, beverages and pharmaceuticals using surfactant modified carbon paste sensor, *Sens. Int.*, 2020, **1**, 100023.
32. T. Miyazawa, M. Itaya, G. C. Burdeos, K. Nakagawa and T. Miyazawa, A critical review of the use of surfactant-coated nanoparticles in nanomedicine and food nanotechnology, *Int. J. Nanomed.*, 2021, **16**, 3937–3999.
33. L. Kvítek, A. Panáček, J. Soukupová, M. Kolář, R. Večeřová and R. Prucek, *et al.*, Effect of surfactants and polymers on stability and antibacterial activity of silver nanoparticles (NPs), *J. Phys. Chem. C*, 2008, **112**(15), 5825–5834. Available from: https://pubs.acs.org/doi/full/10.1021/jp711616v.
34. A. E. Karatapanis, Y. Fiamegos and C. D. Stalikas, Silica-modified magnetic nanoparticles functionalized with cetylpyridinium bromide for the preconcentration of metals after complexation with 8-hydroxyquinoline, *Talanta*, 2011, **84**(3), 834–839.
35. A. Zahid, A. Lashin, U. A. Rana, N. Al-Arifi, I. Ullah and D. D. Dionysiou, *et al.*, Development of surfactant based electrochemical sensor for the trace level detection of mercury, *Electrochim. Acta*, 2016, **190**, 1007–1014.
36. M. Gao, X. Long, J. Du, M. Teng, W. Zhang and Y. Wang, *et al.*, Enhanced curcumin solubility and antibacterial activity by encapsulation in PLGA oily core nanocapsules, *Food Funct.*, 2020, 448–455.
37. Y. Yuan, H. Li, C. Liu, S. Zhang, Y. Xu and D. Wang, Fabrication and Characterization of Lutein-Loaded Nanoparticles Based on Zein and Sophorolipid: Enhancement of Water Solubility, Stability, and Bioaccessibility, *J. Agric. Food Chem.*, 2019, **67**(43), 11977–11985.
38. E. Fortunati, M. Peltzer, I. Armentano, L. Torre, A. Jiménez and J. M. Kenny, Effects of modified cellulose nanocrystals on the barrier and migration properties of PLA nano-biocomposites, *Carbohydr. Polym.*, 2012, **90**(2), 948–956. Available from: https://pubmed.ncbi.nlm.nih.gov/22840025/.
39. F. Mustafa and S. Andreescu, Nanotechnology-based approaches for food sensing and packaging applications, *RSC Adv.*, 2020, **10**(33),

19309–19336. Available from: https://pubs.rsc.org/en/content/articlehtml/2020/ra/d0ra01084g.

40. G. P. Nikoleli, D. P. Nikolelis, C. G. Siontorou, S. Karapetis and T. Varzakas, Novel Biosensors for the Rapid Detection of Toxicants in Foods, *Adv. Food Nutr. Res.*, 2018, **84**, 57–102. Available from: https://pubmed.ncbi.nlm.nih.gov/29555073/.
41. M. Bijad, H. Karimi-Maleh, M. Farsi and S. A. Shahidi, An electrochemical-amplified-platform based on the nanostructure voltammetric sensor for the determination of carmoisine in the presence of tartrazine in dried fruit and soft drink samples, *J. Food Meas. Charact.*, 2018, **12**(1), 634–640. Available from: https://link.springer.com/article/10.1007/s11694-017-9676-1.
42. S. M. Ghoreishi, M. Behpour, N. Jafari and M. Golestaneh, Electrochemical Determination of Tyrosine in the Presence of Dopamine and Uric Acid at the Surface of Gold Nanoparticles Modified Carbon Paste Electrode, *J. Chin. Chem. Soc.*, 2012, **59**(8), 1015–1020. Available from: https://onlinelibrary.wiley.com/doi/full/10.1002/jccs.201100654.
43. J. G. Manjunatha, A surfactant enhanced graphene paste electrode as an effective electrochemical sensor for the sensitive and simultaneous determination of catechol and resorcinol, *Chem. Data Collect.*, 2020, **25**, 100331.
44. C. Raril, J. G. Manjunatha, L. Nanjundaswamy, G. Siddaraju, D. K. Ravishankar and S. Fattepur, *et al.*, Surfactant Immobilized Electrochemical Sensor for the Detection of Indigotine, *Anal. Bioanal. Electrochem.*, 2018, **5**(10), 22368–22375. Available from: www.abechem.com.
45. D. K. Fogliatto, A. M. J. Barbosa and V. S. Ferreira, Voltammetric determination of the phytoestrogen genistein in soy flours and soy-based supplements using cationic surfactant cetyltrimetylammonium bromide, *Colloids Surf., B*, 2010, **78**(2), 243–249.
46. A. S. Farag, M. Sýs, T. Hájek and K. Vytřas, Voltammetric determination of ethylvanillin and methylvanillin sum at carbon paste electrode modified by sodium dodecyl sulfate in selected foodstuffs, *Monatsh. Chem. – Chem. Mon.*, 2018, **149**(11), 1945–1953.
47. A. A. Abdullah, Y. Yardım and Z. Şentürk, The performance of cathodically pretreated boron-doped diamond electrode in cationic surfactant media for enhancing the adsorptive stripping voltammetric determination of catechol-containing flavonoid quercetin in apple juice, *Talanta*, 2018, **187**, 156–164. Available from: https://pubmed.ncbi.nlm.nih.gov/29853029/.
48. M. S. Khan, M. T. Rehman, S. A. Bhat, S. Tabrez, A. Hussain and F. M. Husain, *et al.*, Food additive dye (quinoline yellow) promotes unfolding and aggregation of myoglobin: A spectroscopic and molecular docking analysis, *Spectrochim. Acta, Part A*, 2019, **214**, 216–226.
49. C. Raril, J. G. Manjunatha and G. Tigari, Low-cost voltammetric sensor based on an anionic surfactant modified carbon nanocomposite material for the rapid determination of curcumin in natural food supplement, *Instrum. Sci. Technol.*, 2020, **48**(5), 561–582.

CHAPTER 9

Surfactant Sensors for Pharmaceutical/Medical Applications

S. IREM KAYA,[a] AHMET CETINKAYA[b,c] AND SIBEL A. OZKAN*[b]

[a] Department of Analytical Chemistry, Gulhane Faculty of Pharmacy, University of Health Sciences, Ankara, Turkey; [b] Department of Analytical Chemistry, Faculty of Pharmacy, Ankara University, Ankara, Turkey; [c] Ankara University, Graduate School of Health Sciences, Ankara, Turkey
*Email: sibel.ozkan@pharmacy.ankara.edu.tr

9.1 Introduction

Surfactants, in other words, surface-active agents, are unique, organic compounds that comprise both hydrophilic and hydrophobic parts. What gives surfactants this amphiphilic feature is that the regions defined as the head and tail have different properties. Thanks to this structure, surfactants significantly reduce the interfacial and surface tension between two or more immiscible phases.[1–3] It is possible to group surfactants as cationic, anionic, neutral, or zwitterionic based on the polar part, and there are lots of members of these groups, such as sodium dodecyl sulfate (SDS), cetyltrimethylammonium bromide (CTAB), Triton X-100, cetylpyridinium chloride (CPC), *etc.*[1,2,4,5] Industrial uses of surfactants are particularly prominent. They are used for various purposes in agriculture, food, textile, chemical, cleaning, and pharmaceutical fields.[1,3] Another important use of surfactants is in sensor development studies. A sensor can be defined as a device that converts a physical/chemical/biological change into a measurable signal.

Detection Science Series No. 23
Surfactant-based Sensors in Chemical and Biochemical Detection
Edited by J. G. Manjunatha

Published by the Royal Society of Chemistry, www.rsc.org

Advanced devices are being developed using this basic principle to determine a wide range of analytes.

In sensor studies, drug-active substances can be considered the most popular choice of analytes. Analyses related to the drug processes, such as quality and safety control of drugs and determination in pharmaceutical, environmental, and biological samples, have always been a vital topic for researchers. New technologies and developments in the medical and pharmaceutical industries have contributed to the dramatic increase in the types and number of drugs, along with their prescriptions.[6,7] Additionally, biomarkers or other biological compounds associated with pathological processes of the human body are also significant analytes for sensors. Using sensors to diagnose and monitor diseases and manage treatment protocols are important in medical applications.[8]

Sensor development studies focus on improving sensor performance as a whole. However, when it is evaluated in detail, there are several parameters, such as enhancing sensitivity, stability, accuracy or repeatability, enabling miniaturization and portability, *etc.* In this context, surfactants can have various functions. In chromatographic applications, surfactants whose concentration is above the critical micellar concentration can be included in the mobile phase composition to provide solubility for substances with different charges, hydrophobic properties, and molecular weights. It may also be possible to reduce hazardous solvents in this way and acquire a green analytical chemistry approach.[5,9] Signal response and effective results in electrochemical analysis are directly related to the redox mechanism and electron transfer rate. Therefore, in order to enhance the sensor performance and acquire higher sensitivity, modifications and other outside factors should affect the electron transfer kinetics on the electrode surface. Surfactants can form an adsorption layer on the electrode surface and, thanks to this, it is possible to adjust electrostatic and hydrophilic properties. By this means, higher electron transfer rate, changing redox potential and charge transfer coefficient, and preventing fouling of the electrode surface while working with carbon paste electrodes can be obtained.[2,10,11] Capillary electrophoresis is one of the most important methods used for protein separation. However, a problem that can be frequently encountered in such studies is that proteins adsorb to the capillary and contaminate it. This can cause instability in the electroosmotic flow, which may adversely affect the analysis and separation.[12] In order to overcome this problem, the use of surfactants in the coating of the capillary wall stands out as a cheap, easy-to-use, and renewable option. The ability of surfactants to provide stability and control of electroosmotic flow, and to prevent adsorption of proteins enables effective separation in capillary electrophoresis. For this purpose, several surfactants, such as CTAB, didodecyldimethylammonium bromide (DDAB), and dioctadecyldimethyl ammonium bromide (DODAB) can be employed.[12,13]

Today, sensor applications in the medical and pharmaceutical fields have gained importance and speed. Maximizing the ease of use and commercializing the sensors are also among the aims. Surfactant-based sensors, which provide significant advantages in increasing various aspects of sensor performance, have come to the fore in these analyses in recent years.

It has been seen that electrochemical sensors, which are advantageous with features such as cost-effectiveness, fast and user-friendly analysis, and high sensitivity compared to other methods, stand out, especially surfactant-based sensors.

In this chapter, the use of surfactants in sensor applications will be evaluated thoroughly in terms of electrochemical, chromatographic, and electrophoretic techniques. Additionally, the most recent and prominent studies in medical and pharmaceutical applications of surfactant-based sensors in the last five years will be overviewed, focusing on the electrochemical applications and assessing their analytical parameters. Finally, a critical survey will be done on this subject, and expected developments in the future will be emphasized.

9.2 Surfactants and Their Use in Sensor Applications

Today, there is an increase in the use of surfactants in many industries. Due to the unique structure of surfactants, it is possible to analyze the components using different techniques, such as sample preparation using low amounts of organic solvents, electroanalysis, chromatography, and capillary electrophoresis. In this section, studies on sensor applications using anionic, cationic, and uncharged surfactants will be summarized. The studies of electrochemical surfactant-based sensors, which are the most important part of sensor applications, in pharmaceutical analysis will be overviewed. In addition to these, chromatographic and electrophoresis-based applications of surfactants in the literature will be briefly mentioned.

9.2.1 Electrochemical Sensors

In recent years, studies on the use of surfactants in electroanalytical methods have increased. Surfactants are preferred because of their molecular structure, affecting the behavior of the analytes on the electrode surface and accelerating electron transfer. In this section, studies on the determination of analytes using different electrochemical methods are discussed in detail.

Using a modified electrode with carbon nanohorns (CNH) functionalized with CTAB, an electrochemical bisphenol A (BPA) sensor with excellent sensitivity was fabricated by Zhang *et al.*[14] The carboxylated form of CNH was readily dispersed in water, and the CNH-modified electrode exhibited easy electron transfer and high conductivity. Thanks to the electrostatic interaction, the carboxylic CNH was functionalized with CTAB. The developed CTAB–CNH/GCE sensor was used to detect BPA using differential pulse voltammetry (DPV). Furthermore, the electrostatic interaction and electrocatalytic activity between CTAB and CNH showed a high electrochemical response to the oxidation of BPA. The developed sensor showed a broad linear response to BPA between 0.01 and 30 μL under optimum experimental conditions and a low detection limit of 5.6 nM. The proposed sensor was successfully applied to detect BPA in real samples. The sensor also exhibited strong repeatability and reusability features. This practical and sensitive sensor could also be easily applied to monitor other minor toxic or hazardous chemicals in food and environmental samples.

In another study, Barman *et al.*[15] developed high-performance non-enzymatic glucose sensors using the surfactant-assisted electrochemical technique (SAET) on polished Au thin-film electrodes using highly catalytic platinum nanoparticles (PtNPs). The produced electrodes were immersed in a solution of tetrahydrofuran (THF) to remove any remaining surfactants (Figure 9.1). While the morphological characterization of the electrodes was performed by emission scanning electron microscopy (FESEM), X-ray diffraction (XRD), and energy dispersive spectroscopy (EDX), electrochemical characterizations were made with cyclic voltammetry (CV), and electrochemical impedance spectroscopy (EIS). The production of the biosensor, which has high sensitivity and a wide linearity range for glucose, was obtained after optimization studies, taking into account the deposition potential and time of the nanoparticles. The developed sensor exhibited advantages such as 0.25–36 μM range, a fast reaction time of less than 5 s, high repeatability, and stability. The proposed SAET proved usable for a variety of applications that require uniform deposition and dispersion of other nanoparticles. In conclusion, the proposed platform may be helpful in detecting glucose in environmental and diagnostic bioprocess monitoring.

9.2.2 Chromatographic Techniques

The micellar pseudo phase and the aqueous phase are two distinct media that are present in micellar solutions of surfactants at concentrations above the critical micellar concentration. They may then be utilized as mobile phase components in chromatographic studies due to their strong solubilization ability for compounds of various polarities.[16] Separation is defined as the differential partitioning of analytes between the micelle pseudo phase, bulk mobile phase, and stationary phase. One advantage of micellar liquid chromatography (MLC) over conventional reversed-phase liquid chromatography is the potential role of hydrophilic, hydrophobic, electrostatic, and other interactions in the separation of analytes. The selectivity in this mode can be varied by adjusting the pH, ionic strength, temperature, and concentration of the surfactant and organic modifier fraction in the mobile phase.[5] Moreover, the MLC method is often preferred to identify a wide variety of analytes such as drugs, aromatic carboxylic acids, amino acids, secondary plant metabolites, herbicides, and pesticides.

SDS, inexpensive, readily accessible, biodegradable, and environmental-friendly with respect to the literature, is the surfactant used most often in MLC.[5] It may also denature proteins and has a high solubility in aqueous solutions. In addition, the use of non-ionic surfactants such as Brij-35,[17] Brij L23,[18] and Tween 20[19] in MLC and cationic surfactants such as CTAB[20] has been examined recently.

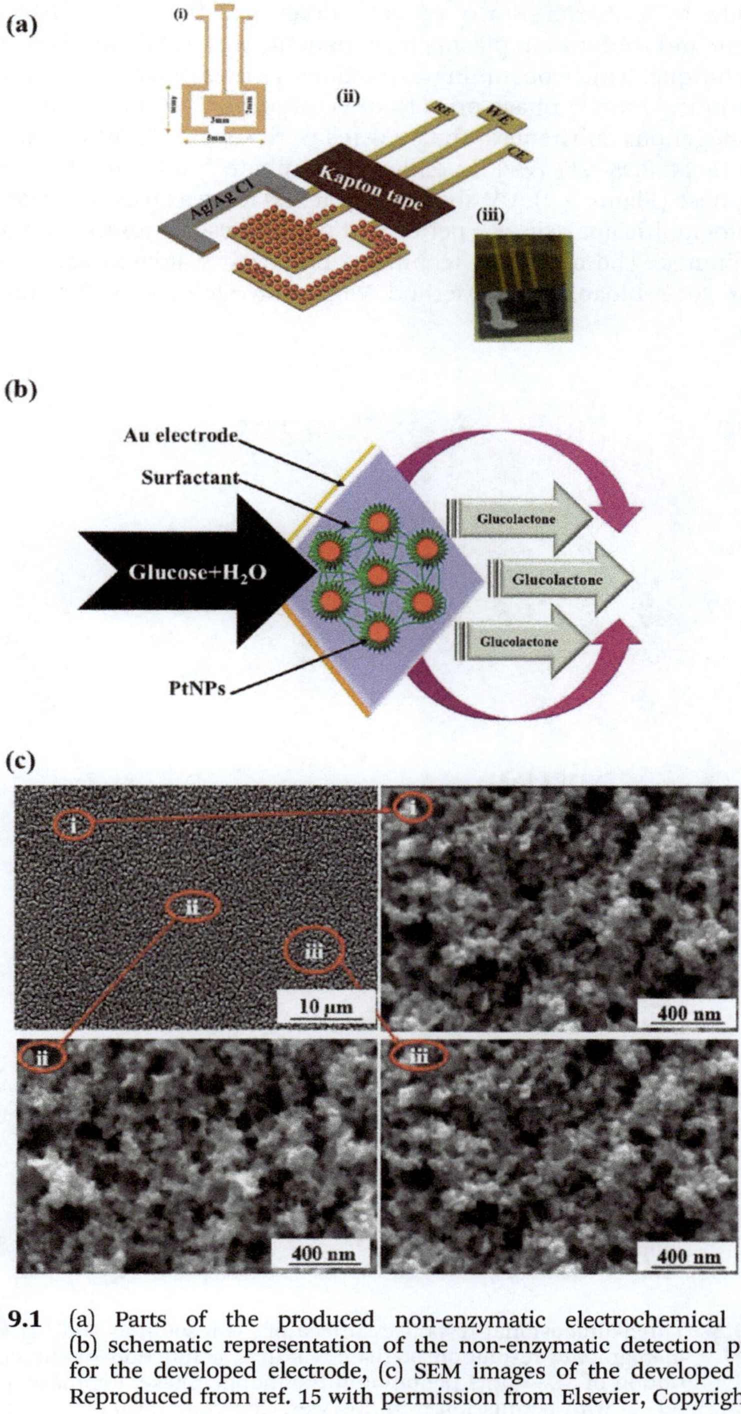

Figure 9.1 (a) Parts of the produced non-enzymatic electrochemical sensor, (b) schematic representation of the non-enzymatic detection principle for the developed electrode, (c) SEM images of the developed sensor. Reproduced from ref. 15 with permission from Elsevier, Copyright 2018.

A study by Goberna Bravo *et al.*[21] described the determination of rifampicin and rifabutin in plasma from patients with tuberculosis using an MLC technique. Time-consuming extractions were avoided by diluting the samples in the mobile phase prior to direct injection. Without interference from endogenous substances, the analytes were separated from the matrix using 0.15 M SDS–6% (v/v) 1-pentanol/phosphate buffer at pH 3 in the mobile phase (Figure 9.2). UV absorbance at 270 nm was used for detection. Chromatographic analysis was performed under these conditions for a total of 22 minutes. The analytical technique has been validated according to the FDA 2018 Bioanalytical Method Validation Guidelines for Industry.

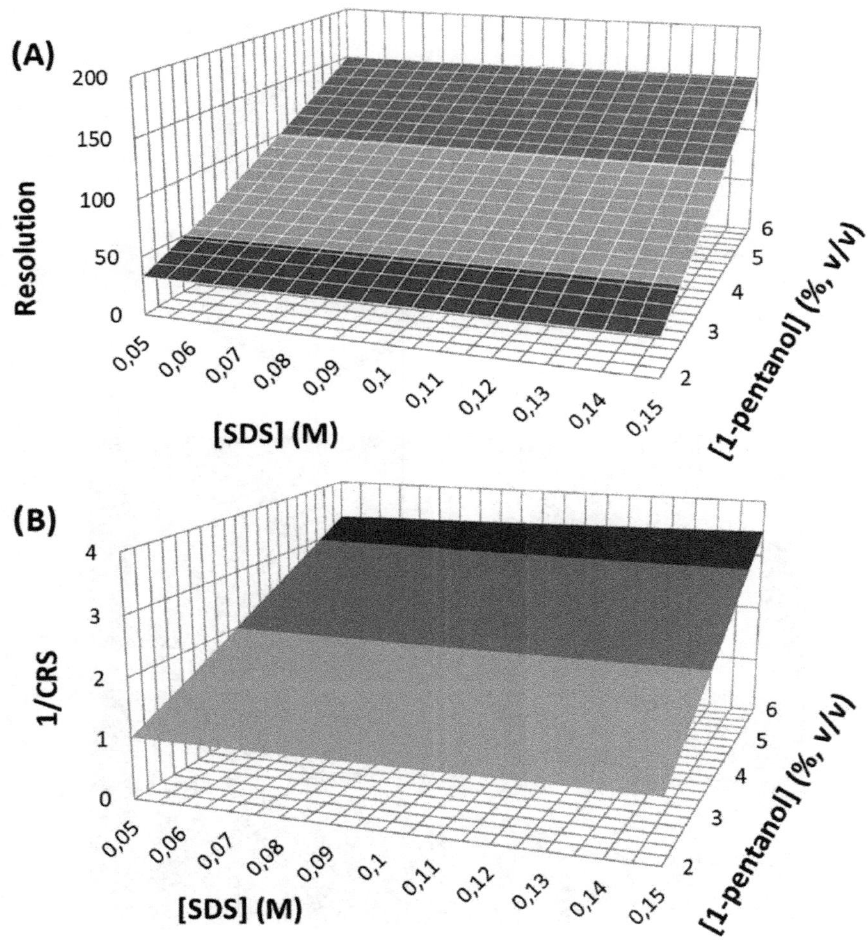

Figure 9.2 Three-dimensional response surface of (A) resolution and (B) 1/chromatographic resolution statistic (CRS) as a function of the concentration of SDS and 1-pentanol in the mobile phase. Reproduced from ref. 21 with permission from Elsevier, Copyright 2020.

The linear range and detection limit were sufficient to find typical rifampicin and rifabutin levels in the patients' plasma. In addition, this approach can be an alternative in clinical laboratories due to its environmental friendliness and high sample yield.

In another study, Ramezani and Absalan[22] successfully used a natural deep eutectic solvent (NADES) and modified MLC in combination with butanol for the analysis of important cardiovascular drugs such as aspirin, atorvastatin, metformin, and metoprolol (Figure 9.3). Primary experimental parameters such as SDS, NADES, butanol, and glacial acetic acid were optimized using multivariate software tools. The ideal volume ratios for SDS (0.09 M), butanol, NADES, and acetic acid were found to be 83 : 10 : 3.5 : 3.5, respectively. As a result, butanol (% v/v) was the variable that had the greatest effect on the resolution between aspirin–metformin ($R_{ASA-MTF}$) response. It is noteworthy that the coefficient signs in this example were negative for butanol and NADES and positive for SDS. In other words, butanol (% v/v) and NADES (% v/v) performed in the opposite direction to SDS in $R_{ASA-MTF}$. This behavior may be caused by the presentation of metformin as a completely positive species according to the distribution of acid–base species, while ASA as a completely negative species under operating conditions (pH 3.5). Consequently, under these conditions, MTF and negatively charged SDS formed a strong relationship through electrostatic contact, while ASA was repelled by SDS. In addition, the robustness of the approach was evaluated using curvature and impact plots. Drugs were completely separated in 12 minutes of total analysis time under ideal circumstances, where the mobile phase flow rate was 1 mL min^{-1}. The Food and Drug Administration (FDA) criteria for bioanalytical analysis were used to validate the performance of the established technique. Drugs in urine and plasma samples have been quantified with success using the intended MLC method.

9.2.3 Capillary Electrophoresis

The basis of the CE separation method is the change in the electrophoretic mobilities of the analytes in an electric field. Key benefits include a simple setup based on load-to-size ratio, high efficiency, short separation time, high resolution, and minimal sample and solvent consumption. There are several forms of CE, including capillary gel electrophoresis, capillary region electrophoresis, and capillary isoelectric focusing.[23] Also, its broad uses include the separation of proteins,[24] drugs,[25] ions,[26] and microbes.[27] This method is frequently used, especially in the analysis of proteins. As a result of increased protein adsorption in fused silica capillaries, the capillary surface changes over time, leading to poor separation and reproducibility. To solve these problems, the researchers produced dynamic, semi-permanent and permanent capillary coatings and were able to overcome them using surfactants.[28]

In the method developed by Hsieh *et al.*,[29] sodium di-(2-ethylhexyl)-sulfosuccinate (AOT) was used in place of conventional anionic surfactants

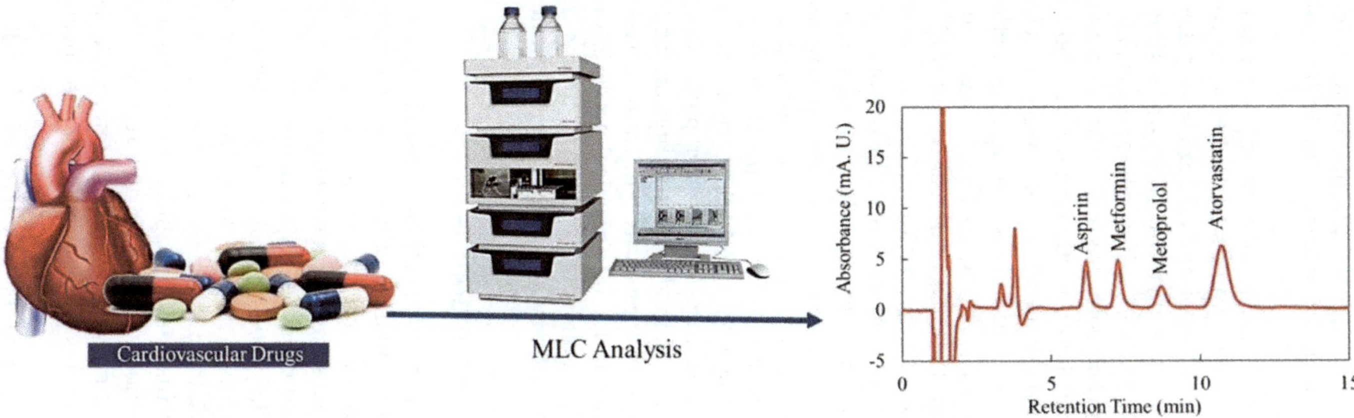

Figure 9.3 The modification of MLC to separate four major cardiovascular drugs of varying polarity. Reproduced from ref. 22 with permission from Elsevier, Copyright 2020.

such as SDS for the simultaneous analysis of eight agonists in animal feeds, using field-amplified sample injection and scavenging (Figure 9.4). Compared to SDS micelles, AOT vesicles offered better β-agonist resolution. By applying this capillary electrophoresis (CE) technique, the detection limits of the eight agonists were above 5 ng mL^{-1}. The stacking CE technique improved sensitivity 400–2000 times compared to the traditional micellar electrokinetic chromatography (MEKC) method (sample injection, 1 psi for 5 s). This approach was successfully used to analyze four animal feeds after the method was validated. It can be said that this method, which has high solubility for eight agonists, is suitable for the regular control of animal feeds.

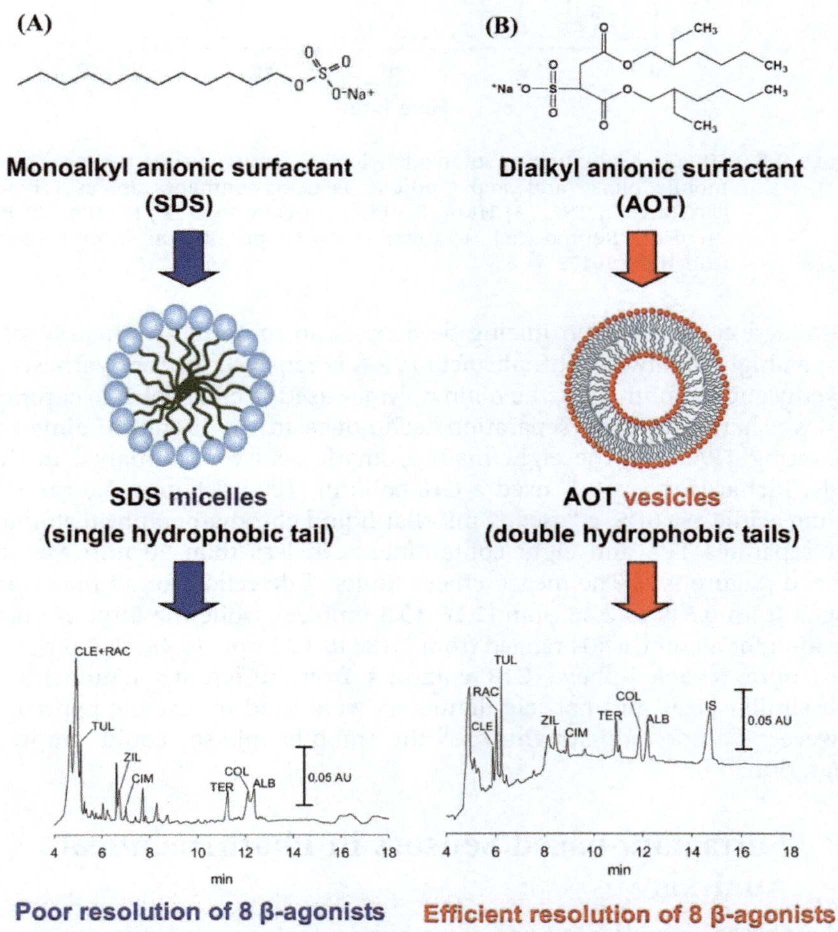

Figure 9.4 The utilization of (A) SDS or (B) AOT in sweeping buffer for analysis of eight β-agonists spiked in animal feeds. Reproduced from ref. 29 with permission from Elsevier, Copyright 2017.

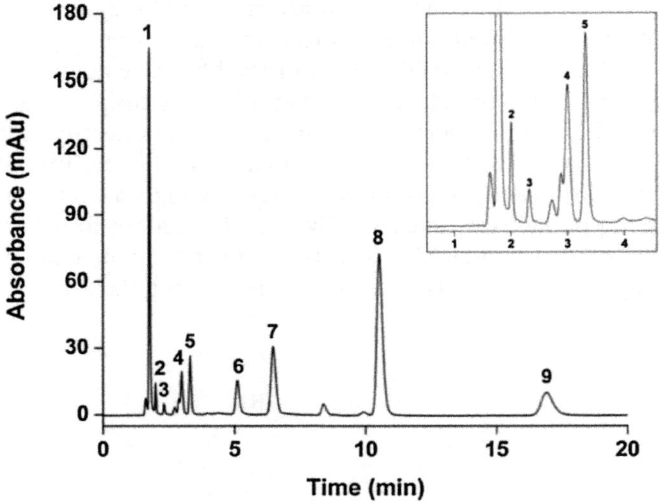

Figure 9.5 HPLC chromatogram of oxidized PTA sample using the optimized mobile phase and step gradient. Peak assignments are as follows: (1) TMA, (2) TSA, (3) HMA, (4) FBA, (5) OPA, (6) TPA, (7) IPA, (8) BA, (9) *p*-TA. Reproduced from ref. 30 with permission from Elsevier, Copyright 2017.

Production of TPA by oxidizing *p*-xylene is an important industrial process, as high-quality terephthalic acid (TPA) is required for the synthesis of polyethylene terephthalate, the main polymer used to create plastic beverage bottles. There are many separation techniques in the literature aimed at separating TPA from the eight main aromatic acid contaminants. In this study, Richardson *et al.*,[30] used a C18 column (100 × 2.1 mm, 3.5 m) containing acidic 1% SDS, a "green" micellar liquid chromatography technique that separates TPA and eight contaminants in less than 20 min was developed (Figure 9.5). The measurement limits of detection for all nine acids ranged from 0.549 to 3.45 ppm (2.16–19.3 pmoles), while the limit of quantification for all nine acids ranged from 0.180 to 1.53 ppm (6.48–43.0 pmoles). Two more reversed-phase C18 columns from different manufacturers with similar sizes and particle diameters were used to test the approach. However, minor adjustments to the mobile phase could improve separation.

9.3 Surfactant-based Sensors in Pharmaceutical Analysis

Various spectroscopic methods are reported for drug analysis, including capillary electrophoresis (CE),[31] high-performance liquid chromatography-ultraviolet (HPLC-UV),[32] HPLC-DAD,[33] high-performance liquid chromatography with fluorescence detection (HPLC-FLD),[34] ultra-high-performance liquid chromatography-diode array detector (UPLC-DAD),[20] up-conversion luminescence performance (UCPL) and UV-vis diffuse reflectance (UV-vis DRS),[35] fluorescence,[36] circular dichroism (CD),[37] and HPLC methods. The disadvantages of these methods are that they are time-consuming and very expensive. On the other hand, electrochemical techniques would be a good alternative source due to their advantages, such as cheap, shorter analysis time, and high sensitivity and selectivity.[38] Recently, surfactant-based sensor applications using different electrodes have been widely used in electrochemical analysis. Table 9.1 summarizes the literature on surfactant-based drug analyses over the past three years.

Due to the unique properties of surfactants, they were investigated for drug analysis in many electrochemical studies and used in combination with different nanomaterials to increase the sensitivity and selectivity of electrochemical sensors. In addition, by assessing their features, such as electrostatic interactions with the sensor surface fabricated in the study and compatibility with the structure of the substance to be examined, surfactants used in electrochemical methods can be chosen from anionic, cationic, or uncharged surfactants. In electrochemical research, SDS, CTAB, and TrX100 are the most often utilized surfactants.

For instance, Sawkar *et al.*[11] developed a GO-SDS/CPE sensor using voltametric methods for the determination of atorvastatin. Atomic force microscopy (AFM) and scanning electron microscopy (SEM) were used to examine the surface topography of graphene oxide, a modifying material. The peak current of atorvastatin at pH 5.0 phosphate buffer solution increased noticeably as a result of the addition of GO and anionic surfactant to the CPE. The linearity range was calculated as 0.81–9.0 μM and the limit of detection (LOD) as 0.038 μM (Figure 9.6). The feasibility of the proposed sensor was tested using pharmaceutical and urine sample analysis. Because of the good recovery values, it could be used for quantitative analysis of pharmaceutical and biological samples. The proposed approach is applicable for the detection and measurement of atorvastatin, as it is reliable and convenient.

In another study by Kesavan *et al.*,[40] the thermal decomposition method (TDM) was effectively used to fabricate and compare the structural, morphological, and electrochemical properties of pristine and surfactant-supported strontium ferrite spinel nanocrystals (Figure 9.7). Surfactant had a significant effect on crystallite size, specific surface area, and electrochemical properties. Notably, SF-CTAB had 13.0 times greater specific surface area than pristine SF-WoS. Pristine SF-WoS showed greater electron transfer properties under the cathodic redox probe solution, but nanocrystals prepared with surfactant exhibited lower conductivity. However, the electrochemical performance of SF-CTAB was superior to that of pristine and other surfactant-based catalysts. The proposed catalyst had a wide linear range from 0.016 to 658.51 μM under optimal experimental conditions. The sensor also had anti-interference capability in addition to its linear range and exhibited a lower detection limit of 0.007 μM. This study showed that

Table 9.1 Latest applications of surfactant-based sensors in pharmaceutical analysis.[a]

Analyte	Sensor	Surfactant	Linear range	LOD	Sample	Recovery (%)	Ref.
Atorvastatin	GO-SDS/CPE	SDS	0.8–9 μM	0.038 μM	Urine samples	94.2–96.6	11
Atorvastatin	GR-CTAB/CPE	CTAB	0.01–0.8 μM	2.46 nM	Pharmaceutical and urine samples	99.66 95–99.66	38
Cocaine Heroin MDMA Cl-PVP Ketamine	SDS/SPE	SDS	1–30 μM 2.5–30 μM 1–30 μM 2.5–30 μM 2.5–30 μM	0.7 μM 1.8 μM 0.9 μM 1.6 μM 1.1 μM	Oral fluid and serum samples	NA	39
Flutamide	SF-CTAB/GCE	CTAB	0.016–658.51 μM	0.007 μM	Serum and tap water samples	95.8–100.16 94.0–101.5	40
Paclitaxel	MWCNT-CTAB/GCE	CTAB	0.1–12 μM	0.438 nM	Injection and urine samples	99.32–201.35 99.07–102.17	41
Natamycin	SDS/BDDE	SDS	0.0975–1.16 μM	0.031 μM	Pharmaceutical and red wine samples	99.48–100.82 96.76–98.95	42
Irinotecan	Poly(CTAB-MWCNT)/PGE	CTAB	2–500 $\mu g\,mL^{-1}$	1.03 $\mu g\,mL^{-1}$	Serum samples	98.2–106	43

Paracetamol Naproxen sodium	(CNCS-NDG)-SDS/GCE	SDS	0.01–90 μM 0.1–60 μM	0.0015 μM 0.0018 μM	Pharmaceutical and serum samples	99.66–100.2 99.80–101 99.80–100.13 99.94–100.3	44
Albendazole	CTAB@g-C_3N_4/MCPE	CTAB	0.2–10 μM	0.01 μM	Pharmaceutical, tap water, urine, and serum samples	99.04–101.09 96.89–98.76 97.26–101.89 97.18–98.97	45
Doxorubicin dacarbazine	SDS/MCPE	SDS	10–80 μM 10–70 μM	1.12 μM 0.82 μM	NA	NA	46
Hydrazine	CSA-PTHA/GCE	CSA	10–90 μM	0.18 μM	NA	NA	47
Ciprofloxacin	PEB-SDS-MCPsE	SDS	18.3–61.1 μM	18.3 μM	Pharmaceutical sample	97.90–100.40	48
Favipiravir	CPT-BDDE	CTAB	0.64 μM–64 nM 0.64 μM–0.13 mM	18 nM 0.15 μM	Pharmaceutical and serum samples	93.1–98.5	49
Dinoterb HCl	CPT-BDDE	CTAB	3.12 μM–31.2 nM	9.16 nM	River water, and soil samples	92–108	50
Amlodipine besylate hydrochlorothiazide	C70	SDS	0.73–2.6 μM 2.4–22 μM	68 nM 0.15 μM	Pharmaceutical and serum samples	106.5 103.2	51
Ethinylestradiol cyproterone acetate	AgNP-SAE	CTAB	0.64–7.8 μM 1–10 μM	0.103 μM 0.299 μM	Pharmaceutical and urine samples	97.1–107.2	52

Table 9.1 (*Continued*)

Analyte	Sensor	Surfactant	Linear range	LOD	Sample	Recovery (%)	Ref.
Norfloxacin	APT-BDDE	SDS	0.157–12.5 μM	40.7 nM	Urine samples	92	53
Eltrombopag	BDDE	CTAB	0.23–28 μM	57 nM	Pharmaceutical samples	97.3–108	54
Enrofloxacin	APT-BDDE	SDS	2.78 μM–69.6 nM	16.8 nM	Pharmaceutical and urine samples	104.67	55
Ranitidine	PDA/CPsE	SDS CTAB	0.1–7.5 μM	19 nM	Pharmaceutical and urine samples	98–101.3	56

[a] NA: not available, SDS: sodium dodecyl sulfate, CTAB: cetyltrimethylammonium bromide, CAS: camphor sulfonic acid, GO: graphene oxide, GR: graphene, SF: strontium ferrite nanocrystals, MWCNT: multi-walled carbon nanotube, CNCS: carboxylated nanocellulose, NDG: nitrogen-doped graphene, g-C_3N_4: nitrogen-rich nanosheet, PEB: poly(Evans blue), PTHA: polythiophene/aluminium oxide, C70: 70% graphite, CPT: cathodically pretreated, APT: anodically pretreated, PDA: poly(dopamine), CPE: carbon paste electrode, SPE: screen-printed electrode, GCE: glassy carbon electrode, BDDE: boron-doped diamond electrode, MCPE: modified carbon paste electrode, PGE: pencil graphite electrode, MCPsE: doubly modified carbon paste electrode, AgNP-SAE: silver nanoparticle solid amalgam electrode.

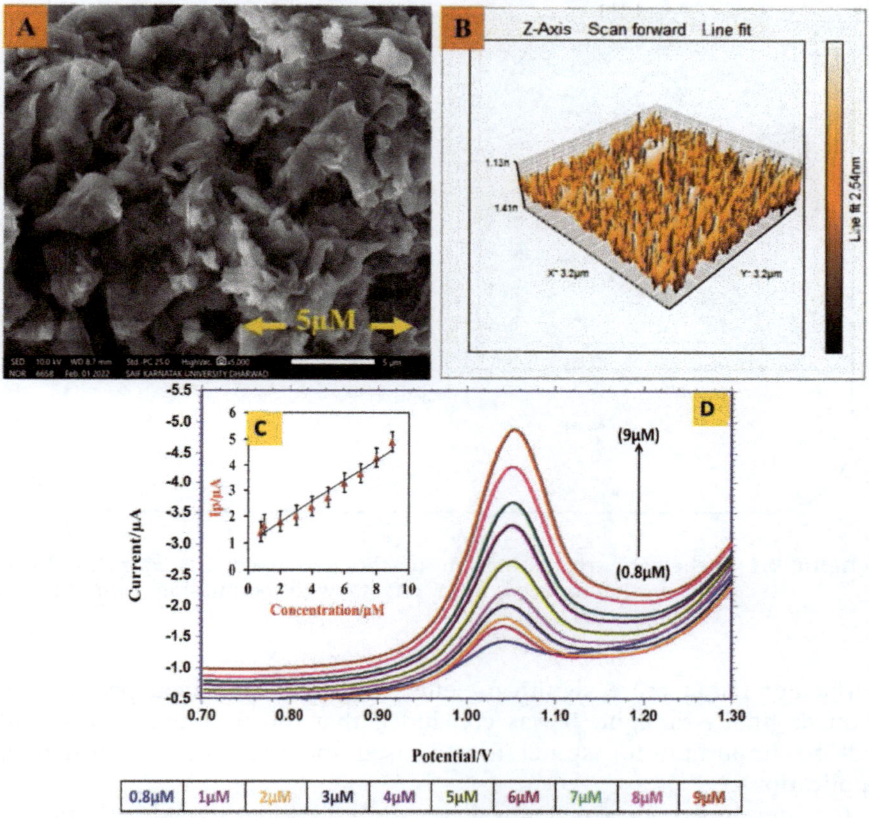

Figure 9.6 (A) SEM image, (B) AFM image for GO, (C) plot of I_p *vs.* concentration, (D) DPV at a GO-SDS/CPE. Reproduced from ref. 11 with permission from Elsevier, Copyright 2022.

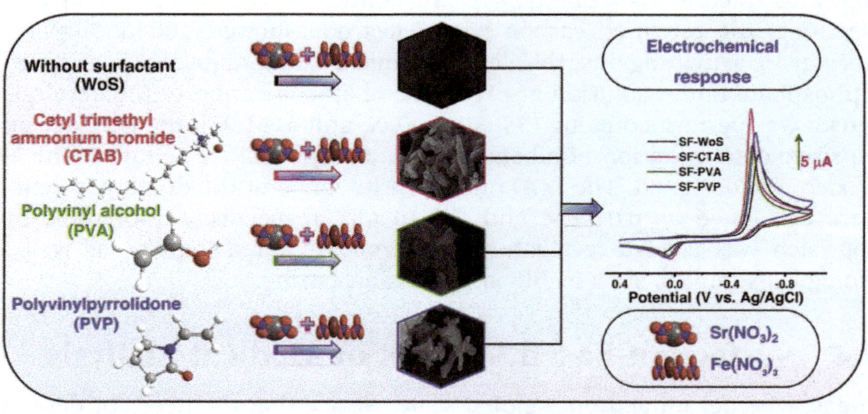

Figure 9.7 Synthesis of strontium ferrite nanocrystals. Reproduced from ref. 40 with permission from Elsevier, Copyright 2022.

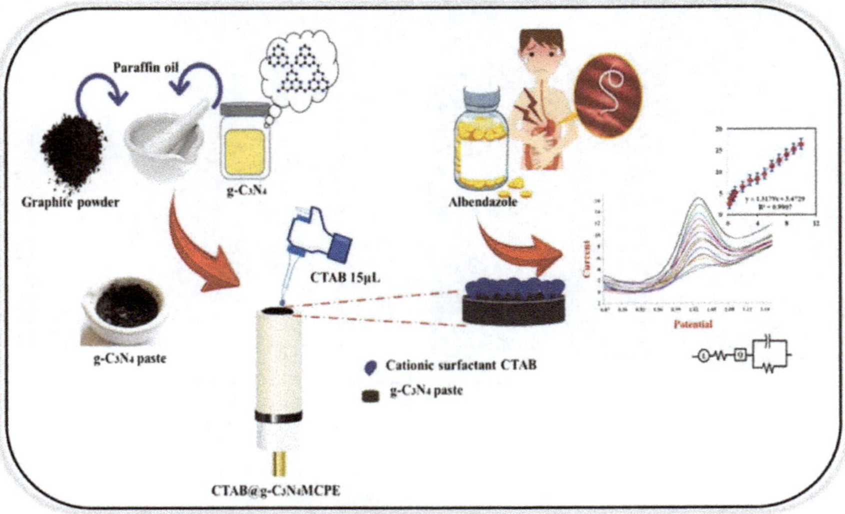

Scheme 9.1 Schematic representation of the developed CTAB@g-C_3N_4–MCPE sensor. Reproduced from ref. 45 with permission from Elsevier, Copyright 2022.

surfactant has a more significant effect on electrochemical performance than originally thought. It was concluded that the findings of this study will be important for spinel ferrite-based materials in electrochemical applications.

The developed electrochemical sensor CTAB@g-C_3N_4–MCPE was developed for the determination of albendazoline by Patil *et al.*[45] In this study, CTAB, a single-chain cationic surfactant fixed on the g-C_3N_4–MCPE surface, was used to create an electrochemical sensor for the detection of albendazoline at trace levels (Scheme 9.1). In contrast to the bare carbon paste electrode, the modified carbon paste electrode showed outstanding electrocatalytic activity against the electrochemical oxidation of albendazoline in a phosphate buffer solution at pH 3. The characterization of the developed sensor was performed using EIS, SEM, EDX, and AFM. For the accurate and sensitive determination of albendazoline, a very sensitive voltammetric approach was designed. The LOD and linearity range of the developed sensor were calculated as 0.01 μM and 0.2–10 μM, respectively. Moreover, this approach was used to evaluate albendazole in water samples as well as biological samples, such as blood plasma and urine.

9.4 Surfactant-based Sensors in Medical Analysis

Today, when examined on a global scale, we see that the need for developments and advances in the health field is increasing. In this context, the prevention, diagnosis, and treatment of diseases are discussed within the scope of medical applications. Various diagnostic devices and sensors used for this purpose shed light on early diagnosis and appropriate treatment with the determination of different biomarkers. In addition, the analysis of blood levels of various biological materials and metabolites is effective not only in the diagnosis stage but also in the follow-up of the prognosis of the disease and the effectiveness of the treatment procedure. The aim is to develop sensor devices to be used in medical analyses for purposes such as early diagnosis, effective treatment, and increasing life expectancy and quality for patients, with certain features: selectivity, high sensitivity, ease of application, low cost, biocompatibility, flexibility, and portability. In this way, it is possible to gain socioeconomic and medical advantages.[8,57–59]

The ability of surfactants to change some features on the electrode surface, thanks to their surface-active agent properties, has made them important materials in electrochemical sensors. The ions of the surfactant create an adsorption layer on the electrode surface and improve electrochemical properties, such as intensity of the electrochemical peaks and redox potential, and enhance the stability of the electrochemical response, resulting in increased sensitivity and lower LOD values.[10,11,60] Other significant roles of surfactants are used in base material and morphological mediators, decreasing interfacial tension between liquid phases. These properties are used to obtain stable and uniform metal–organic frameworks (MOFs) in various sizes and shapes. Due to their higher surface area and good electrical conductivity, MOFs are used in electrochemical sensor applications. MOFs with improved features using surfactants are preferred to increase sensor performance.[61] Surfactants can also be employed in the contents of the mobile phase in chromatographic techniques. They increase the solubility of different solutions with various polarity properties. Thus, it becomes easier for the analytes to be separated.[62] Another application of surfactants is that they can be utilized to modify the molecular orientation of liquid crystals that are used as novel sensing platforms for the detection of a wide variety of compounds, ranging from proteins and enzymes to nucleic acids and metabolites. The changes in molecular orientations by surfactants are suitable for optical detection, leading to sensor applications in various fields.[63]

All the advantageous features of surfactants listed above contribute to the achievement of goals such as selectivity, high sensitivity, ease of application, and flexibility, which are aimed to be achieved in medical sensors, and therefore surfactants are frequently preferred in sensor applications. The most recent and prominent studies on the development of surfactant-based sensors for the detection of various analytes that may be useful in medical applications are compiled in Table 9.2. These studies were examined in terms of selected surfactant, linear range, LOD, analyzed sample and recovery percentage.

For the determination of an important neurotransmitter in the central nervous system, serotonin, and the essential amino acid L-tryptophan, which is a precursor in serotonin synthesis, an electrochemical method was

Table 9.2 Latest applications of surfactant-based sensors in medical analysis.[a]

Analyte	Sensor	Surfactant	Linear range	LOD	Sample	Recovery (%)	Ref.
Serotonin L-Tryptophan	CTAB/NiS/CS/GCE	CTAB	0.01–200.0 μmol L^{-1}	3.35 nmol L^{-1} 2.94 nmol L^{-1}	Serum, urine and injection samples	94.5–111.5	60
Rutin	SDSMGPE	SDS	2.0–10 μM 10–60 μM	1.17×10^{-8} M	Fruit juice	93.82–98.32	64
Glucose	Chronoamperometric sensor based on CuO nanostructures	TRX, CTAB, PEG	3.0 μM–3.0 mM	NA	NA	NA	65
Catalase	H_2O_2-responsive LC sensing platform	$PBEC_{14}A$	5–75 mU mL^{-1}	5.5 mU mL^{-1}	Human serum	NA	66
Trypsin	Label-free fluorometric assay	EDAB	5–150 ng mL^{-1}	4.2 ng mL^{-1}	Human urine	96.18–103.2	67
Cholesterol	AuNPs-f-MWCNT-PPy-ChOx/GCE	TRX	2×10^{-3}–8×10^{-3} M	0.1×10^{-3} M	Human serum	93.5–127.6	68
Carboxylesterase	Surfactant-doped LC sensor	OTB	18–180 U L^{-1}	18 U L^{-1}	Human urine	NA	63
Dopamine	Ru(II) complex-SDSB luminescent probe	SDBS	0.1–1.0 μM 2.0–10.0 μM	6.6 nM	NA	NA	69

Bilirubin	Visual detection-based fluorescence sensor	Amphiphilic polystyrene	1×10^{-7}–1×10^{-5} M	20 nM	Human serum	NA	70
Ascorbic acid	TOAB/ERGO@YD/GCE	TOAB	1.33 μM–1.46 mM	0.28 μM	Vitamin C tablets, kiwi fruit, soft drinks, human urine	96–106	71
Glucose Methanol	NiOOH/EPGCE	PVP	0.5–117 μM	0.2 μM	NA	NA	72
HSA	Fluorescent sensor	SDS	0–67.0 μg mL^{-1}	3.48 ng L^{-1}	Pond water and wastewater	99.9–101.7	73

[a] CTAB: cetyltrimethylammonium bromide, NiS/CS: NiS/carbon hybrids with NiS nanoparticles embedded on 3D hollow porous carbon microspheres, GCE: glassy carbon electrode, SDS: sodium dodecyl sulfate, SDSMGPE: sodium dodecyl sulfate chemically modified graphene paste electrode, TRX: Triton X-100, PEG: polyethylene glycol, NA: not available, $PBEC_{14}A$: *N*,*N*-dimethyl-*N*-(4-(4,4,5,5-tetramethyl-1,3,2-dioxaborolan-2-yl)benzyl)tetradecan-1-aminium bromide, LC: liquid crystal, EDAB: ethylene-bis(dodecyl dimethyl ammonium bromide), AuNPs-*f*-MWCNT: gold nanoparticles functionalized multiwalled carbon nanotube, PPy: polypyrrole, ChOx: cholesterol oxidase, OTB: *N*-octadecyloxycarbonylmethyl-*N*,*N*,*N*-trimethylammonium bromide, SDBS: sodium dodecyl benzyl sulfonate, TOAB: tetraoctylammonium bromide, ERGO: electrochemically reduced graphene oxide, YD: zinc porphyrin dye YD2-*o*-C8, PVP: polyvinyl pyrrolidone, NiOOH: aqueous NiOOH nanosheets, EPGCE: electrochemically pretreated GCE, HAS: human serum albumin, VOC: volatile organic compounds, DOC: sodium deoxycholate.

developed by Dong and Zhang.[60] These two molecules have important bodily functions in connection with each other, and problems are seen in their deficiencies. For example, decreases in serotonin levels due to L-tryptophan deficiency can result in depression and obsession. Therefore, the development of fast, selective, and sensitive sensors for the determination of serotonin and L-tryptophan from biological and pharmaceutical samples is important in the field of medicine, especially in diagnostics. In this study, the electrochemical analysis method was preferred because it is very advantageous in terms of sensitivity and rapidness. In addition, while the sensor was being developed, it was aimed to increase stability by using the cationic surfactant CTAB, and to increase the conductivity and active surface area by using 3D NiS/CS hollow microspheres (Scheme 9.2). CTAB has a role as the membrane matrix, and additionally, it provides enrichment of analytes using hydrophobic interactions. As a result, thanks to the synergistic effect obtained, high sensitivity and low LOD values (3.35 and 2.94 nmol L^{-1}) for serotonin and L-tryptophan were achieved. The results of the recovery study using the serum and urine of mice as real samples were found to be compatible with HPLC. It was stated that the developed sensor is effective and sensitive and can be used in the biomedical field for the determination of serotonin and L-tryptophan.

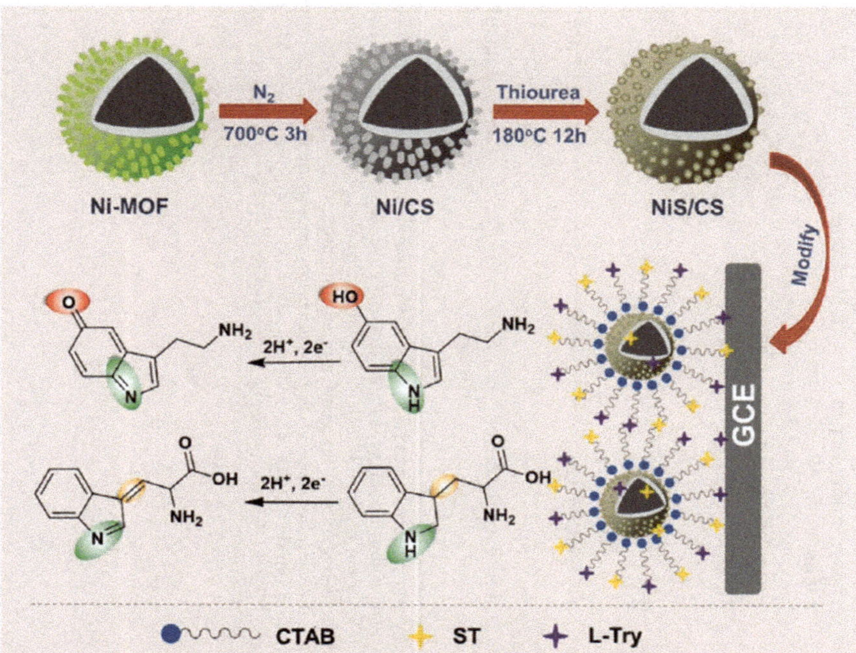

Scheme 9.2 The preparation steps for 3D NiS/CS hollow microspheres. Reproduced from ref. 60 with permission from Elsevier, Copyright 2022.

Glucose determination has a critical place in the management of diabetes because it is the easiest way for patients to acquire a medical test by themselves outside of a hospital environment. Therefore, the development of sensors with improved sensitivity is still a hot topic for researchers due to its medical importance. Patil *et al.*[65] developed a non-enzymatic chronoamperometric glucose sensor using CuO thin films. CuO nanostructures are advantageous materials in electrochemistry due to their non-hazardous profile and ability to improve electron transfer rate. Using various surfactants in the development process of nanostructures enables the control and modification of morphologies. For this purpose, in this study, the effects of CTAB, polyethylene glycol (PEG), and triton X-100 (TRX) surfactants in solving problems in CuO nanostructure synthesis and sensor efficiency were investigated. When the sensitivity was interpreted by evaluating the current values in chronoamperometric studies, it was seen that the best result was obtained with the CuO:TRX electrode. It was also found that the results obtained with different surfactants were better than for electrodes prepared without surfactants. With this study, it was demonstrated that surfactants could improve electrochemical properties and performance through their effects on morphology.

Trypsin is an important digestive enzyme secreted from the pancreas and is a specific biomarker used in the evaluation of pancreatic function. Pathological phenomena in the pancreas can be assessed through changes in trypsin level, and in this respect, it is often preferred in diagnostics as a reliable biomarker. Therefore, it is important to develop effective and sensitive sensors for the accurate and precise determination of trypsin. Label-free fluorometric probe assays are valuable tools for trypsin detection. Supramolecular assemblies formed with surfactants have been used recently, especially for fluorometric enzyme assays. However, it is difficult to work in physiological environments because they are affected by high salt concentrations. Gemini surfactants, on the other hand, show stability at high salt concentrations. In this study by Yuan *et al.*,[67] ethylene-bis(dodecyl dimethyl ammonium bromide) (EDAB), which is a gemini surfactant, was used in the composition of a label-free fluorescence turn-on trypsin assay to form a supramolecular assembly. The linear range was found to be between 5.0 and 150.0 $ng\,mL^{-1}$, with an LOD value of 4.2 $ng\,mL^{-1}$ for trypsin determination. In studies with urine as a real sample, trypsin was not found in healthy volunteers, and studies with spiking yielded good relative standard deviation % (RSD%) results with recovery values between 95.8% and 103.2%. This demonstrated the applicability of the sensor and its potential for use in clinical diagnostics.

Carboxylesterase (CES) is an enzyme that is responsible for the catalytic hydrolysis of various substances, which takes place in detoxification and metabolic activation processes. Due to its prominent liver-related effects, it is an important biomarker for hepatocellular carcinoma. The development of reliable and effective sensors for the detection and monitoring of CES has great significance for advantageous diagnostic tools for hepatocellular carcinoma in medical applications. For this purpose, liquid crystal (LC)-based

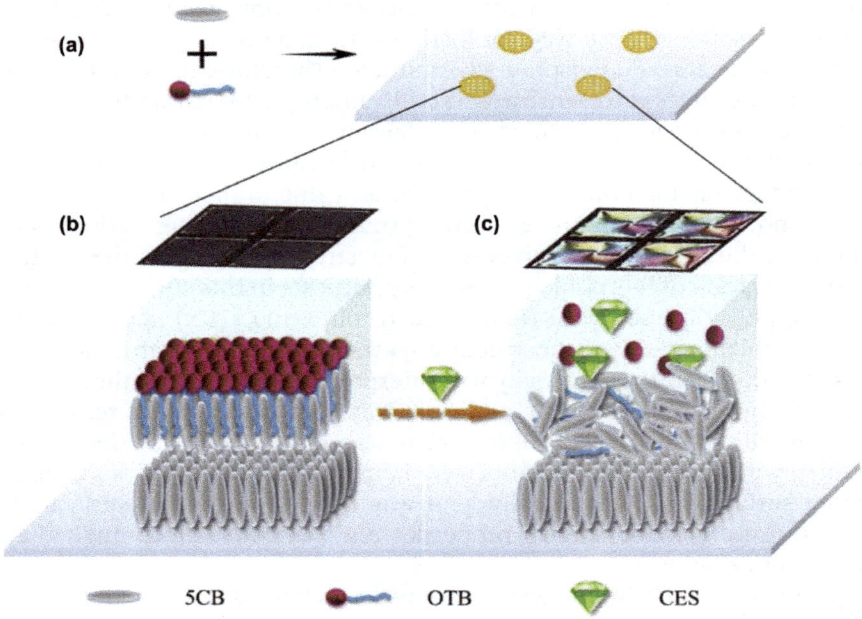

Figure 9.8 Development process for an LC-based sensor for CES. Reproduced from ref. 63 with permission from Elsevier, Copyright 2019.

sensors can be used as advantageous platforms that have the ability to transform biological and chemical signals into optical images that can be observed with the naked eye. Zhou *et al.*[63] synthesized a novel surfactant, *N*-octadecyloxy carbonylmethyl-*N*,*N*,*N*-trimethylammonium bromide (OTB), and used it in the development of an LC-based sensor for CES (Figure 9.8). The mechanism of the developed sensor is based on the degradation of an OTB monolayer by CES hydrolyzing the OTB, followed by differences in the optical responses of the LC/aqueous interface. Even though this newly developed sensor has a close LOD value (18 $U L^{-1}$) to other available methods in the literature, it stands out with several advantages, such as being label-free, rapid, low-cost, and easy to use. It also has good selectivity in the presence of other enzymes and biological compounds such as glucose, uric acid, lipase, sulfatase, *etc.* It can be said that this novel LC-based platform has the potential for applications in the medical field.

9.5 Future Outlook

As described in detail in the previous parts of the chapter, surfactant-based sensors provide several advantages in medical and pharmaceutical applications. However, the common focus of today's sensor work – commercialization, conversion to portable and miniature devices – applies to surfactant-based sensors as well. Thus, it will be possible to enable point-of-care applications for patients that provide faster and easier analysis. Having fast sensors that patients can use without going to the hospital can bring great advantages, especially for critical patients who constantly need to monitor the markers and parameters associated with disease. When evaluated in terms of pharmaceutical analysis, the same aims also apply. The application area and samples for drug analysis are broad in scope; for example, while analyzing environmental samples, the miniaturization and portability that allows the analysis to be performed without a laboratory environment can be a significant feature. Besides, commercial rapid sensor devices for drug analysis are a specific need for biological samples collected from patients.

Today, another important focus in analytical sensor development studies is the reduction/elimination of environmentally harmful parameters of the analysis and the choice of a green approach. When integrating this green approach with surfactant-based sensors, a new approach, biologically-based "green" surfactants, can be used. Various microorganisms such as fungi, bacteria, and yeasts that can produce biosurfactants are reported in several studies. In addition, some plants can contain compounds with amphiphilic features, such as saponins, lecithin, and phospholipids. All of these bio-based green surfactants can be used in food, cosmetics, pharmaceuticals, and agricultural applications as biocompatible, non-toxic, and natural alternatives.

9.6 Conclusion

When studies in recent years are examined, it is possible to say that surfactants have started to take a critical place in sensor applications thanks to the many advantages they provide. When we evaluate pharmaceutical sensor applications using surfactants, electrochemical sensors stand out with several advantages, such as rapid and affordable analysis, and improved sensitivity and selectivity compared to other available techniques. In order to increase the performance of electrochemical sensors, surfactants improve the electron transfer rate on the electrode surface and affect the sensitivity. Anionic surfactant SDS and cationic surfactant CTAB are the most preferred in drug determination studies conducted in recent years. The electrodes used differ according to the type of analysis, such as GCE, BDDE, or CPE. It can be said that the techniques used in the medical applications of surfactant-based sensors vary. Likewise, there is a wide range of preferred surfactants, such as OTB, TRX, PEG, ETAB, *etc.* Medical sensors are also mostly used for diagnostic purposes, and they are used to determine guiding compounds related to pathological processes in the body, such as enzymes and markers.

As a result, medical and pharmaceutical applications of surfactant-based sensors are among the rapidly developing and increasingly important areas, and the advantageous properties of different sensors continue to be a popular topic for researchers.

Acknowledgements

Ahmet Cetinkaya thanks the financial support from the Council of Higher Education 100/2000 (YOK) under the special 100/2000 and the TUBITAK under the BIDEB/2211-A Ph.D. and ARDEB/1004 Scholarship Programmes.

References

1. M. Aguirre-Ramírez, H. Silva-Jiménez, I. M. Banat and M. A. Díaz De Rienzo, Surfactants: physicochemical interactions with biological macromolecules, *Biotechnol. Lett.*, 2021, **43**, 523–535.
2. M. R. Housaindokht, F. Janati-Fard and N. Ashraf, Recent advances in applications of surfactant-based voltammetric sensors, *J. Surfactants Deterg.*, 2021, **24**, 873–895.
3. S. M. Shaban, J. Kang and D. H. Kim, Surfactants: Recent advances and their applications, *Compos. Commun.*, 2020, **22**, 100537.
4. N. Sakač, D. Marković, B. Šarkanj, D. Madunić-Čačić, K. Hajdek, B. Smoljan and M. Jozanović, Direct potentiometric study of cationic and nonionic surfactants in disinfectants and personal care products by new surfactant sensor based on 1,3-dihexadecyl-1H-benzo[d]imidazol-3-ium, *Molecules*, 2021, **26**, 1–16.
5. J. Esteve-Romero, J. Albiol-Chiva and J. Peris-Vicente, A review on development of analytical methods to determine monitorable drugs in serum and urine by micellar liquid chromatography using direct injection, *Anal. Chim. Acta*, 2016, **926**, 1–16.
6. S. Chen, Z. Li, S. Zhang, Y. Zhou, X. Xiao, P. Cui, B. Xu, Q. Zhao, S. Kong and Y. Dai, Emerging biotechnology applications in natural product and synthetic pharmaceutical analyses, *Acta Pharm. Sin. B*, 2022, **12**, 4075–4097.
7. B. Nigović, S. Jurić and A. Mornar, Electrochemical determination of nepafenac topically applied nonsteroidal anti-inflammatory drug using graphene nanoplatelets-carbon nanofibers modified glassy carbon electrode, *J. Electroanal. Chem.*, 2018, **817**, 30–35.
8. A. A. Ensafi, Z. Saberi and N. Kazemifard, in *Functionalized Nanomaterial-Based Electrochemical Sensors Principles, Fabrication Methods, and Applications*, ed. C. M. Hussain and J. G. Manjunatha, Woodhead Publishing, 2022, pp. 277–308.
9. K. M. Wilson and N. D. Danielson, Micellar and sub-micellar chromatography with a cocamidopropyl betaine surfactant, *J. Chromatogr. A*, 2022, **1681**, 463442.
10. V. B. Patil, S. J. Malode, S. M. Tuwar and N. P. Shetti, Graphene sheet-based electrochemical sensor with cationic surfactant for sensitive detection of atorvastatin, *Sensors Int.*, 2022, **3**, 100198.
11. R. R. Sawkar, V. B. Patil and S. M. Tuwar, Electrochemical oxidation of Atorvastatin using graphene oxide and surfactant-based sensor, *Mater. Today Proc.*, 2023, **74**, 267–273.

12. Y. Tian, Y. Li, J. Mei, B. Cai, J. Dong, Z. Shi and Y. Xiao, Simultaneous separation of acidic and basic proteins using gemini pyrrolidinium surfactants and hexafluoroisopropanol as dynamic coating additives in capillary electrophoresis, *J. Chromatogr. A*, 2015, **1412**, 151–158.
13. L. Shulman, L. Pei, M. F. Bahnasy and C. A. Lucy, High pH instability of quaternary ammonium surfactant coatings in capillary electrophoresis, *Analyst*, 2017, **142**, 2145–2151.
14. J. Zhang, X. Xu and Z. Chen, A highly sensitive electrochemical sensor for bisphenol A using cetyltrimethylammonium bromide functionalized carbon nanohorn modified electrode, *Ionics*, 2018, **24**, 2123–2134.
15. S. C. Barman, M. F. Hossain and J. Y. Park, Soft surfactant-assisted uniformly dispersed platinum nanoparticles for high performance electrochemical non-enzymatic glucose sensing platform, *J. Electroanal. Chem.*, 2018, **824**, 121–127.
16. D. W. Armstrong and S. J. Henry, Use of an aqueous micellar mobile phase for separation of phenols and polynuclear aromatic hydrocarbons via HPLC, *J. Liq. Chromatogr.*, 1980, **3**, 657–662.
17. Y. M. Dong, N. Li, Q. An and N. W. Lu, A novel nonionic micellar liquid chromatographic method for simultaneous determination of pseudo-ephedrine, paracetamol, and chlorpheniramine in cold compound preparations, *J. Liq. Chromatogr. Relat. Technol.*, 2015, **38**, 251–258.
18. B. Otašević, J. Šljivić, A. Protić, N. Maljurić, A. Malenović and M. Zečević, Comparison of AQbD and grid point search methodology in the development of micellar HPLC method for the analysis of cilazapril and hydrochlorothiazide dosage form stability, *Microchem. J.*, 2019, **145**, 655–663.
19. A. A. K. F. Ali and N. D. Danielson, Micellar and sub-micellar liquid chromatography of terephthalic acid contaminants using a C18 column coated with Tween 20, *Anal. Chim. Acta*, 2020, **1105**, 214–223.
20. H. M. Albishri, A. M. Almalawi, W. Alshitari and D. A. El-Hady, Cyclodextrin-Modified Micellar UPLC for Direct, Sensitive and Selective Determination of Water Soluble Vitamins in Milk, *J. Chromatogr. Sci.*, 2020, **58**, 203–210.
21. M. Á. G. Bravo, A. Durgbanshi, D. Bose, P. Mishra, J. Albiol-Chiva, J. Esteve-Romero and J. Peris-Vicente, Quantification of rifampicin and rifabutin in plasma of tuberculosis patients by micellar liquid chromatography, *Microchem. J.*, 2020, **157**, 104865.
22. A. M. Ramezani and G. Absalan, Employment of a natural deep eutectic solvent as a sustainable mobile phase additive for improving the isolation of four crucial cardiovascular drugs by micellar liquid chromatography, *J. Pharm. Biomed. Anal.*, 2020, **186**, 113259.
23. Z. Gao and W. Zhong, Recent (2018–2020) development in capillary electrophoresis, *Anal. Bioanal. Chem.*, 2022, **414**, 115–130.
24. A. Shala-Lawrence, M. J. Blanden, S. M. Krylova, S. A. Gangopadhyay, S. S. Beloborodov, J. L. Hougland and S. N. Krylov, Simultaneous Analysis of a Non-Lipidated Protein and Its Lipidated Counterpart:

Enabling Quantitative Investigation of Protein Lipidation's Impact on Cellular Regulation, *Anal. Chem.*, 2017, **89**, 13502–13507.
25. M. Hernández-Mesa, C. Cruces-Blanco and A. M. García-Campaña, Capillary electrophoresis-tandem mass spectrometry combined with molecularly imprinted solid phase extraction as useful tool for the monitoring of 5-nitroimidazoles and their metabolites in urine samples, *Talanta*, 2017, **163**, 111–120.
26. K. L. Chen, S. J. Jiang and Y. L. Chen, Determining lead, cadmium and mercury in cosmetics using sweeping via dynamic chelation by capillary electrophoresis, *Anal. Bioanal. Chem.*, 2017, **409**, 2461–2469.
27. M. Horká, P. Karásek, M. Roth and F. Růžička, Pre-concentration and separation of bacteria by volume coupling electrophoresis on supercritical water-etched fused silica capillary with two segments of different internal diameters and inner surface roughnesses, *Anal. Bioanal. Chem.*, 2018, **410**, 167–175.
28. C. L. Crihfield, C. J. Kristoff, L. M. Veltri, W. M. Penny and L. A. Holland, Semi-permanent cationic coating for protein separations, *J. Chromatogr. A*, 2019, **1607**, 460397.
29. S. Y. Hsieh, C. C. Wang, H. S. Kou and S. M. Wu, Dialkyl anionic surfactant in field-amplified sample injection and sweeping-micellar electrokinetic chromatography for determination of eight leanness-promoting β-agonists in animal feeds, *J. Pharm. Biomed. Anal.*, 2017, **141**, 222–228.
30. A. E. Richardson, S. D. McPherson, J. M. Fasciano, R. E. Pauls and N. D. Danielson, Micellar liquid chromatography of terephthalic acid impurities, *J. Chromatogr. A*, 2017, **1491**, 67–74.
31. Ş. Stăcescu, G. Hancu, L. Gagyi, R. M. Soare and H. Kelemen, Simultaneous determination of hydrochlorothiazide and telmisartan from pharmaceutical preparations using capillary electrophoresis, *Stud. Univ. Babes-Bolyai, Chem.*, 2017, **62**, 189–198.
32. M. Hadjmohammadi and M. Salary, Application of sigmoidal transformation functions in optimization of micellar liquid chromatographic separation of six quinolone antibiotics, *J. Chromatogr. Sci.*, 2016, **54**, 429–435.
33. I. Garrido-Cano, A. García-García, J. Peris-Vicente, E. Ochoa-Aranda and J. Esteve-Romero, A method to quantify several tyrosine kinase inhibitors in plasma by micellar liquid chromatography and validation according to the European Medicines Agency guidelines, *Talanta*, 2015, **144**, 1287–1295.
34. M. M. Mabrouk, S. F. Hammad, S. F. El-Malla and E. A. Elshenawy, Green micellar HPLC-fluorescence method for simultaneous determination of metoprolol and amlodipine in their combined dosage form: Application on metoprolol in spiked human plasma, *Microchem. J.*, 2019, **147**, 635–642.
35. X. Wang, H. Jing, Y. Gao, Y. Xin and Q. Li, CTAB assisted hydrothermal synthesis of 0D/2D carbon quantum dots $BiOIO_3$ composite structure for

photocatalytic degradation of tetracycline, *Inorg. Chem. Commun.*, 2022, **146**, 110209.

36. B. Zhao, Y. Li, Y. Zhao, Y. Ma, F. Li, H. Han, N. Wang and X. Wang, A sensing platform based on zinc-porphyrin derinative in hexadecyl trimethyl ammonium bromide (CTAB) microemulsion for highly sensitive detection of theophylline, *Spectrochim. Acta – Part A Mol. Biomol. Spectrosc.*, 2022, **281**, 121592.
37. X. Jiang, M. Cui, W. Zhang, L. Xie and L. Xu, Design of chiral mesoporous silica nanorods using ursodeoxycholic acid/chenodeoxycholic acid and CTAB as templates for chiral-selective release of achiral drugs, *Mater. Lett.*, 2021, **285**, 129144.
38. V. B. Patil, S. J. Malode, S. M. Tuwar and N. P. Shetti, Graphene sheet-based electrochemical sensor with cationic surfactant for sensitive detection of atorvastatin, *Sensors Int.*, 2022, **3**, 100198.
39. M. Parrilla, F. Joosten and K. De Wael, Enhanced electrochemical detection of illicit drugs in oral fluid by the use of surfactant-mediated solution, *Sensors Actuators B Chem.*, 2021, **348**, 130659.
40. G. Kesavan, M. Pichumani, S. M. Chen and C. S. Ko, Surfactant-assisted (CTAB, PVA, PVP) thermal decomposition synthesis of strontium spinel ferrite nanocrystals for electrochemical sensing of cytostatic drug flutamide, *Mater. Today Chem.*, 2022, **26**, 101045.
41. J. I. Gowda, R. M. Hanabaratti and S. M. Tuwar, MWCNT modified glassy carbon electrode in presence of cationic surfactant for the electroanalysis of paclitaxel, *Res. Chem.*, 2021, **3**, 100243.
42. K. Mielech-Łukasiewicz and M. Leoniuk, Voltammetric determination of natamycin using a cathodically pretreated boron-doped diamond electrode in the presence of sodium dodecyl sulfate, *Microchem. J.*, 2020, **159**, 105570.
43. G. Bolat, Investigation of poly(CTAB-MWCNTs) composite based electrochemical DNA biosensor and interaction study with anticancer drug Irinotecan, *Microchem. J.*, 2020, **159**, 105426.
44. M. Shalauddin, S. Akhter, W. J. Basirun, N. S. Anuar, O. Akbarzadeh, M. A. Mohammed and M. R. Johan, Carboxylated nanocellulose dispersed nitrogen doped graphene nanosheets and sodium dodecyl sulfate modified electrochemical sensor for the simultaneous determination of paracetamol and naproxen sodium, *Measurement*, 2022, **194**, 110961.
45. Y. N. Patil, M. B. Megalamani and S. T. Nandibewoor, Highly sensitive electro-oxidative voltammetric determination of anthelmintic drug albendazole using porous graphitic carbon nitride sensor infused with cationic micellar solution, *J. Pharm. Biomed. Anal.*, 2022, **221**, 115072.
46. S. Deepa, B. E. K. Swamy and K. V. Pai, A surfactant SDS modified carbon paste electrode as an enhanced and effective electrochemical sensor for the determination of doxorubicin and dacarbazine its applications: A voltammetric study, *J. Electroanal. Chem.*, 2020, **879**, 114748.

47. H. Vijeth, S. P. Ashokkumar, L. Yesappa, M. Vandana and H. Devendrappa, Camphor sulfonic acid surfactant assisted polythiophene nanocomposite for efficient electrochemical hydrazine sensor, *Mater. Res. Express*, 2019, **6**, 125375.
48. P. A. Pushpanjali, J. G. Manjunatha and M. T. Shreenivas, The Electrochemical Resolution of Ciprofloxacin, Riboflavin and Estriol Using Anionic Surfactant and Polymer-Modified Carbon Paste Electrode, *ChemistrySelect*, 2019, **4**, 13427–13433.
49. S. Allahverdiyeva, O. Yunusoğlu, Y. Yardım and Z. Şentürk, First electrochemical evaluation of favipiravir used as an antiviral option in the treatment of COVID-19: A study of its enhanced voltammetric determination in cationic surfactant media using a boron-doped diamond electrode, *Anal. Chim. Acta*, 2021, **1159**, 238418.
50. P. T. Pınar, S. Allahverdiyeva, Y. Yardım and Z. Şentürk, Voltammetric sensing of dinitrophenolic herbicide dinoterb on cathodically pretreated boron-doped diamond electrode in the presence of cationic surfactant, *Microchem. J.*, 2020, **155**, 104772.
51. E. M. da Silva, G. C. de Oliveira, A. B. de Siqueira, A. J. Terezo and M. Castilho, Development of a composite electrode based on graphite and polycaprolactone for the determination of antihypertensive drugs, *Microchem. J.*, 2020, **158**, 105228.
52. V. O. de Jesus, V. S. Ferreira and B. G. Lucca, Electrochemical study and simultaneous voltammetric determination of contraceptives ethinylestradiol and cyproterone acetate using silver nanoparticles solid amalgam electrode and cationic surfactant, *Talanta*, 2020, **210**, 120610.
53. F. Karahan, Z. Başı, E. Keskin, P. T. Pınar, Y. Yardım and Z. Şentürk, Electrochemical Determination of Fluoroquinolone Antibiotic Norfloxacin in the Presence of Anionic Surfactant Using the Anodically Pretreated Boron-Doped Diamond Electrode, *ChemistrySelect*, 2020, **5**, 12862–12868.
54. S. Allahverdiyeva, M. Çalışır, E. Keskin, Y. Yardım and Z. Şentürk, First electroanalytical investigation and simple quantification of a thrombopoietin-receptor agonist drug eltrombopag in the presence of cationic surfactant at a non-modified boron-doped diamond electrode, *Diam. Relat. Mater.*, 2020, **110**, 108146.
55. F. Dönmez, Y. Yardım and Z. Şentürk, Electroanalytical determination of enrofloxacin based on the enhancement effect of the anionic surfactant at anodically pretreated boron-doped diamond electrode, *Diam. Relat. Mater.*, 2018, **84**, 95–102.
56. P. Talay Pınar, Y. Yardım and Z. Şentürk, Electrochemical oxidation of ranitidine at poly(dopamine) modified carbon paste electrode: Its voltammetric determination in pharmaceutical and biological samples based on the enhancement effect of anionic surfactant, *Sens. Actuators, B*, 2018, **273**, 1463–1473.
57. C. I. L. Justino, A. C. Duarte and T. A. P. Rocha-Santos, Critical overview on the application of sensors and biosensors for clinical analysis, *TrAC – Trends Anal. Chem.*, 2016, **85**, 36–60.

58. K. Nemčeková and J. Labuda, Advanced materials-integrated electrochemical sensors as promising medical diagnostics tools: A review, *Mater. Sci. Eng. C*, 2021, **120**, 11751.
59. A. Tricoli and R. Bo, in *Frontiers of Nanoscience*, ed. P. Milani and M. Sowwan, Elsevier, 2020, vol. 15, pp. 247–269.
60. Y. Dong and L. Zhang, Coupling surfactants with 3D hollow raspberry-like NiS/carbon microsphere with S vacancies for enhanced sensitivity monitoring of serotonin and L-tryptophan, *Sens. Actuators, B*, 2022, **368**, 132140.
61. P. Arul, S. T. Huang and V. Mani, Surfactant-induced morphological evolution of Cu(II) metal organic frameworks: Applicable in picomolar quantification of bilirubin, *Appl. Surf. Sci.*, 2021, **557**, 149827.
62. D. N. Unal, S. Yıldırım, S. Kurbanoglu and B. Uslu, Current trends and roles of surfactants for chromatographic and electrochemical sensing, *TrAC, Trends Anal. Chem.*, DOI: 10.1016/j.trac.2021.116418.
63. L. Zhou, Q. Kang, M. Fang and L. Yu, Label-free, rapid, and sensitive detection of carboxylesterase using surfactant-doped liquid crystal sensor, *J. Mol. Liq.*, 2019, **296**, 111921.
64. N. Sreeharsha, M. Telsang, G. M. Basavarajappa and S. Pund, Development of simple and accurate voltammetric method based on Surfactant Modified Graphene for the Determination of Rutin, *Int. J. Electrochem. Sci.*, 2022, **17**, 220228.
65. A. S. Patil, R. T. Patil and V. J. Fulari, Surfactant assisted synthesis of CuO nanostructures for nonenzymatic glucose sensor, *Mater. Today Proc.*, 2021, **46**, 2340–2346.
66. S. Lu, Y. Guo, L. Qi, Q. Hu and L. Yu, Highly sensitive and label-free detection of catalase by a H2O2-responsive liquid crystal sensing platform, *Sens. Actuators, B*, 2021, **344**, 130279.
67. N. Yuan, L. Jia and J. Zhu, Label-free Fluorescence Turn on Trypsin Assay Based on Gemini Surfactant/heparin/Nile Red Supramolecular Assembly, *J. Fluoresc.*, 2021, **31**, 1537–1545.
68. M. Alagappan, S. Immanuel, R. Sivasubramanian and A. Kandaswamy, Development of cholesterol biosensor using Au nanoparticles decorated f-MWCNT covered with polypyrrole network, *Arab. J. Chem.*, 2020, **13**, 2001–2010.
69. S. Sheth, M. Li and Q. Song, New luminescent probe for the selective detection of dopamine based on in situ prepared Ru(II) complex-sodium dodecyl benzyl sulfonate assembly, *J. Photochem. Photobiol., A*, 2019, **371**, 128–135.
70. S. K. Makkad and S. K. Asha, Designed Amphiphilic Polystyrene as Surfactant for Oligo(*p*-phenylenevinylene)-Incorporated PS Nanobeads and Visual Detection of Bilirubin in Human Blood Serum, *ACS Appl. Polym. Mater.*, 2019, **1**, 1230–1239.
71. H. Wu, X. Li, M. Chen, C. Wang, T. Wei, H. Zhang and S. Fan, A nanohybrid based on porphyrin dye functionalized graphene oxide for the application in non-enzymatic electrochemical sensor, *Electrochim. Acta*, 2018, **259**, 355–364.

72. S. J. Li, W. Guo, B. Q. Yuan, D. J. Zhang, Z. Q. Feng and J. M. Du, Assembly of ultrathin NiOOH nanosheets on electrochemically pretreated glassy carbon electrode for electrocatalytic oxidation of glucose and methanol, *Sens. Actuators, B*, 2017, **240**, 398–407.
73. C. Liao, F. Li, S. Huang, B. Zheng, J. Du and D. Xiao, A specific and biocompatible fluorescent sensor based on the hybrid of GFP chromophore and peptide for HSA detection, *Biosens. Bioelectron.*, 2016, **86**, 489–495.

CHAPTER 10

Surfactant-based Ion-selective Electrodes

VINAYAK ADIMULE (0000-0002-2198-6601),*[a]
SANTOSH NANDI (0000-0001-8609-2222),[b]
NIDHI MANHAS (0000-0002-6049-2145),[c]
CHINNA BATHULA (0000-0001-5460-799X)[d] AND
PARASHURAM LAKSHMINARAYANA (0000-0002-3711-2095)[e]

[a] Angadi Institute of Technology and Management (AITM), Savagaon Road, Belagavi-590009, Karnataka, India; [b] KLE Technological University, Dr. M. S. Sheshgiri College of Engineering and Technology, Udyambagh, Belagavi-590008, India, Email: santosh.nandi603@gmail.com; [c] School of Sciences, Indira Gandhi National Open University, Maidan Grahi, New-Delhi, India, Email: nidhimanhas17@gmail.com; [d] Division of Electronics and Electrical Engineering, Dongguk University, Seoul, 04620, Republic of Korea, Email: cdbathula@dongguk.edu; [e] Department of Chemistry, Nitte Meenakshi Institute of Technology, Yelahanka, Bangalore-560064, Karnataka, India, Email: ramacedamy1990@gmail.com
*Emails: adimulevinayak@yahoo.in; researcher@vinayakadimule.in

10.1 Introduction

In recent years, anionic surfactants have commonly been used in the industrial and domestic fields, household agents, and personal care products. Traditionally, measurement of surfactant concentration has required tedious work, *i.e.*, using liquid or gas chromatography or using large amount of solvents (methylene blue method).[1,2] Ion-selective electrodes or ion-selective field effect transistors[3,4] are alternative methods for the estimation of surfactant

Detection Science Series No. 23
Surfactant-based Sensors in Chemical and Biochemical Detection
Edited by J. G. Manjunatha

Published by the Royal Society of Chemistry, www.rsc.org

concentration. Ion selective-electrodes incorporated into potentiometric determination have found enormous application due to low experimental cost, selectivity, and sensitivity, and can be applied to a wide range of experimental methods with variable conditions.[5] Potentiometric titration involving surfactants commonly occurs *via* the charged ions of the surfactants, forming water-insoluble products (salt).[6,7] The electrode potential changes as a result of reaction between equal molar ratios of ionic surfactants can be easily detected at the end point. The literature reports the detection of anionic surfactants by surfactant electrodes using potentiometric methods. Single-walled carbon nanotubes (SWCNTs) have been constructed in order to determine hexadecyltrimethylammonium ion (CTA^+) and sodium dodecyl ion (SD^-) concentrations and reported by Najafi *et al.*[8] Juan Soto *et al.*, constructed ion-selective electrodes for the effective determination of aza-oxa cycloalkanes and cyclam derivatives using polyvinyl chloride (PVC) membrane electrodes. CNT dispersion can be improved by means of physical and chemical methods.[9,10] The dispersion of the polymer chain depends on its ability to form complexes. In the case of surfactants, the dispersion depends on the interaction between the CNTs and the surfactant. The mechanism may be suggested in the following ways: (i) cylindrical micelle formation at the CNT surface,[11] (ii) random surfactant adsorption,[12,13] (iii) adsorption of hemimicelles onto the CNTs.[14] The dispersion of surfactants is closely related to the molecular structure.[15,16] Single-tailed surfactants are poorer dispersants than Gemini surfactants.[17,18] Organic functional groups present in surfactants are better dispersion agents.[19,20] Also, mixed systems of surfactants, *i.e.*, cationic and anionic, possess better dispersion than the individual surfactants.[21,22] The stable and synergistic effects of high concentrations of CNT dispersion has led to their application in ion-selective electrodes.

In the present chapter, the authors describe the feasibility, stability, selectivity of PVC membrane electrodes for the detection of common surfactants like OTAB, HTAB, *etc.* The ion selectivity and sensitivity offered by the electrode, and the mechanism of dispersion of surfactants are discussed. Also, the quantification, mechanism of dispersion of single-chain surfactants over SWCNTs, CNTs, and SWCNTs-NH_2 and understanding of the adsorption of ionic surfactants and nanostructure (NS) are discussed in detail.

10.1.1 Materials and Characterization

All the chemicals and solvents were purchased from Sigma-Aldrich, S-d fine chemicals, e-Merck, *etc.* and used without any further purification. SDS, CTAB, di-*n*-octyl phthalate (DNOT), tetrahydrofuran (THF) were purchased from Sigma-Aldrich India Ltd. Ag/AgCl electrodes were used as inner reference electrodes and saturated calomel electrodes were used as external reference electrodes. Potentiometric determination of the surfactants was carried out using CHI instruments from Shimadzhu. Morphology, microstructure of the adsorbed surfactants and desorbed surfactants were analyzed using a TESCAN VEGA 3 LMU scanning electron microscope. Zeta potentials were measured using a Nicomp Nano Z3000 Zeta Potential Analyzer.

10.1.2 Construction and Preparation of the Electrodes

Surfactant-based ion-selective electrodes were prepared as per the literature-reported procedure.[23] From the obtained membrane a 1.0 cm disc was made, and attached to the PVC tube in THF solution. The inner solution of KCl and SDS was prepared at 1.0×10^{-3} mol L^{-1}. The electrode was preconditioned by immersing in 10^{-3} mol L^{-1} SDS solution.

10.1.3 Preparation of the Ion Pairs and Membranes

The ion pairs (CTA^+/SD^-) were prepared by pouring hot solutions of (1 : 1) sodium dodecyl sulfate and hexadecyl-trimethylammonium bromide (HTAC). White coloured precipitate was obtained, washed with hot water, washed with hot acetone solution, and dried.

10.1.4 Apparatus Involved

Potentiometric determinations were carried out using a pH 3C/mV meter and a saturated calomel electrode (SCE) was used as the external reference electrode. The indicator electrode was a self-made anionic-surfactant electrode. The potentiometric titrations were carried out using an automatic titrant at 25 °C with water thermostat system.

10.2 Experimental

10.2.1 Potentiometric Titrations and Electromotive Force (emf) Measurements

A self-made ion-selective electrode used for all emf measurements. The potentiometer involved the following apparatus or assemblies: inner solution, inner reference electrode (Ag/AgCl), SCE (external reference electrode), PVC membrane, and saturated KCl solution. Selectivity coefficients were determined using pH 3C pH/mV potentiometer using a fixed interference method with an ion concentration of 2.5×10^{-2} mmol L^{-1}. A plot of potential (E) *vs.* log SDS concentration gave a calibration curve. Standard solutions of SDS were prepared in the range 1.0×10^{-7} to 2.5×10^{-2} mol L^{-1}. The test solution used was 25 mL at the rate of 0.1 mL s^{-1}. The concentration of the hymene was 4.043×10^{-3} mol L^{-1}. The all-potentiometric titrations were carried out at 25 °C.

10.3 Results and Discussion

10.3.1 CNTs and Surfactant Dispersion Process

One of the essential applications of CNTs is to form uniform dispersions with ionic surfactants. Previous literature reports show the dispersion of

SWCNT-COOH with ionic surfactants, which is mainly due to the electrostatic interaction of CNTs with ionic surfactant complexes.[20] The best dispersing system is chosen, which is directly dependent on the tail of the surfactant that produces the highest charge on the CNTs. This is attributed to the incremental charge on the CNTs that results from electrostatic repulsion between ionic parts of the surfactants. The present chapter discusses the overall increase in the surface charge for both anionic (SDS) and cationic (Do-TAB) surfactants with respect to SWCNTs and multi-walled carbon nanotubes (MWCNTs) which are non-functionalized, as displayed in Figure 10.1. As the surfactant remains same, the dispersion also remains same for these additives. Positive charge will develop as a result of the dispersion of positively charged CNTs ($SWCNTs-NH_2$, pH = 3) for cationic surfactants, as displayed in Figure 10.2. In case of amine-functionalized SWCNTs, an extra charge value appeared depending on the type and amount of surfactant added. This could be confirmed by the value of zeta potentials

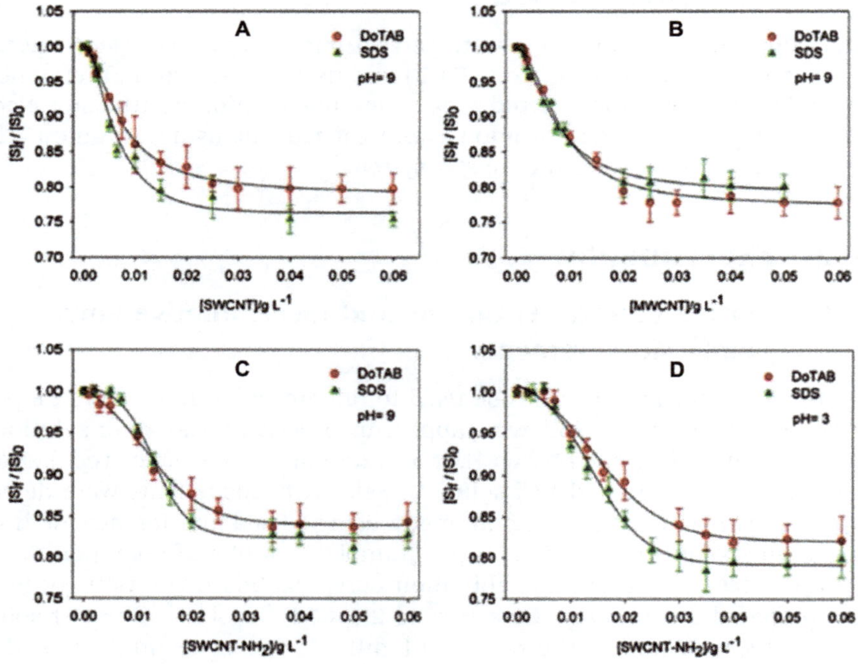

Figure 10.1 Plots of the relative free dodecyltrimethyl ammonium bromide (DoTAB) and sodium dodecyl sulfate (SDS) concentrations *vs.* the carbon nanotube concentration in buffered aqueous solutions: (A) neutral single-walled carbon nanotubes (SWCNT), (B) neutral multi-walled carbon nanotubes (MWCNT), (C) neutral $SWCNT-NH_2$ (pH = 9), and (D) positively charged $SWCNT-NH_2$ (pH = 3). Reproduced from ref. 24, https://doi.org/10.3390/ijms22020826, under the terms of the CC BY 4.0 License, http://creativecommons.org/licenses/by/4.0/.

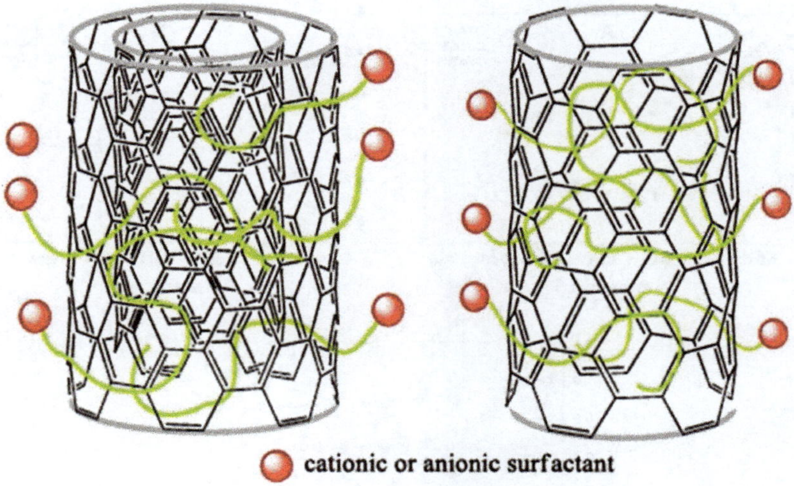

Figure 10.2 Schematic picture of the adsorption of anionic and cationic surfactants on an MWCNT (A) and an SWCNT (B) surface. Reproduced from ref. 24, https://doi.org/10.3390/ijms22020826, under the terms of the CC BY 4.0 License, http://creativecommons.org/licenses/by/4.0/.

recorded in the absence of any surfactants, as well as a decrease in the SDS concentration and an increase in the DoTAB concentration. Therefore, the dispersion process of the nanotubes gave better results for the cationic surfactants.

10.3.2 Surfactant and CNT Adsorption Process

In order to study the adsorption process of surfactants on CNTs, the adsorption of SWCNTs, MWCNTs, and SWCNTs-NH_2 were studied using potentiometric methods using ion-selective electrodes. Previous literature reports have suggested possible interactions of SWCNTs-COOH.[24] The obtained emf gives information about the amount of CNTs that do not interact with the surfactant. The factors that influence the emf of the surfactant activity are given by the Nernst eqn (10.1).

$$E = E^{0\prime} - \frac{RT}{Z_i\, F} \log a_i \tag{10.1}$$

For dilute surfactant concentrations given by $[S]_f$ and slope values close to 59 V at 298 K, a plot of emf *vs.* $\log[S]_f$ gives the free surfactant concentration in the solution. The relative decrease in the free surfactant concentration can be calculated by $[S]_f/[S]_0$, where $[S]_0$ is the total surfactant concentration, as displayed in Figure 10.3. The values indicate that binding of the surfactant to a CNT, followed by binding to another tube is the most favourable condition

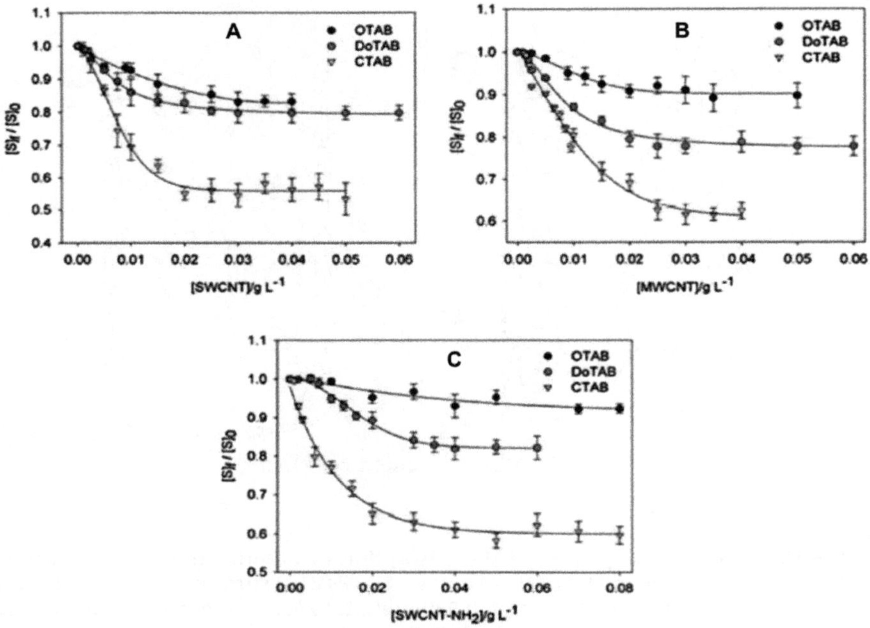

Figure 10.3 Plots of the relative free surfactant concentration *vs.* the carbon nanotube concentration in buffered aqueous solutions: (A) SWCNT (pH = 9), (B) MWCNT (pH = 9), and (C) SWCNT-NH_2 (pH = 3). Reproduced from ref. 24, https://doi.org/10.3390/ijms22020826, under the terms of the CC BY 4.0 License, http://creativecommons.org/licenses/by/4.0/.

for adsorption processes to occur. This has led to experimental[25] and simulation studies.[26] Figure 10.4 displays the CNT–surfactant interaction, which confirms that adsorption occurs mainly on the outer walls of MWCNTs. The surface charge possessed by neutral (SWCNTs, MWCNTs, SWCNTs-NH_2, pH = 9) or positive (SWCNTs-NH_2, pH = 3) nanotubes, and the polar head group of the charge do not influence the adsorption process. Electrostatic interactions do not play avital role in the adsorption of surfactants on SWCNTs.[27] The mechanism of interaction occurs through the hydrophobic tails of the surfactants with the hydrophilic heads directed outwards to the solvent molecules.[28,29] The adsorption process depends on the length of the hydrocarbon chain. Figure 10.5 displays three cationic interactions of surfactants with CNTs, with different chain lengths and same head group. For all the surfactant–CNT complexes, decreases in the concentration of $[S]_f/[S]_0$ values have been investigated. The ratio $[S]_f/[S]_0$ is strongly dependent on the length of the surfactants. As the length of the hydrocarbon chain increases, there is a relative decrease in $[S]_f/[S]_0$. Figure 10.5 displays CNT–surfactant complexes with charges that influence their formation. Figure 10.5 displays critical miceller concentration (cmc),

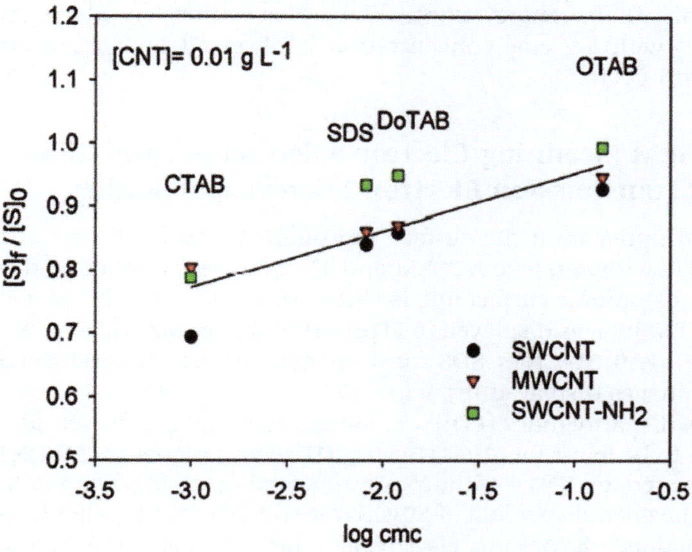

Figure 10.4 Plot of $[S]_f/[S]_0$, at $[CNT] = 0.01\ g\,L^{-1}$, *versus* $\log(cmc\ mol^{-1}\,dm^{-3})$ for all ionic surfactants and CNTs investigated. Reproduced from ref. 24, https://doi.org/10.3390/ijms22020826, under the terms of the CC BY 4.0 License, http://creativecommons.org/licenses/by/4.0/.

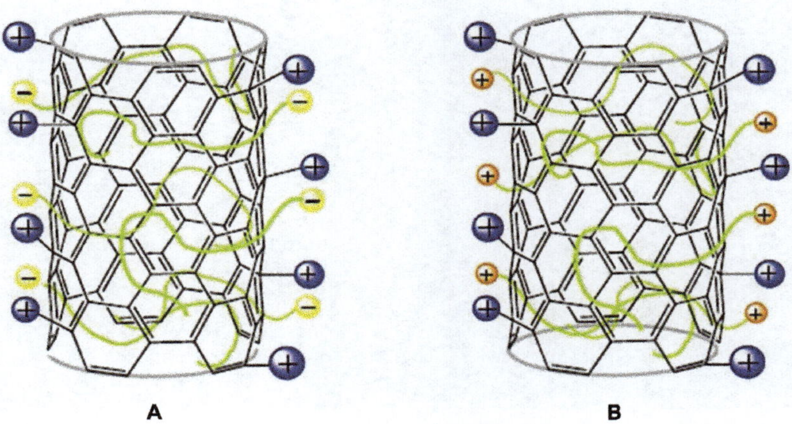

Figure 10.5 Schematic picture of the adsorption of anionic (A) and cationic (B) surfactants on the charged SWCNT-NH_2 surfaces (pH = 3). Reproduced from ref. 24, https://doi.org/10.3390/ijms22020826, under the terms of the CC BY 4.0 License, http://creativecommons.org/licenses/by/4.0/

hydrophobic Gibbs energy curve. Good correlation of $[S]_f/[S]_0$ *vs.* CNT (0.01 $g L^{-1}$) with log cmc concentration is observed for all the ionic surfactants with CNTs.[30]

10.3.3 SEM (Scanning Electron Microscope) and TEM (Transmission Electron Microscope) Studies

In order to understand the surface morphology and grain size of the dispersed CNTs with surfactants, SEM and TEM analysis were carried out. The dispersion of cationic surfactants is better with CNTs and can be confirmed from the TEM images displayed in Figure 10.6. All the sizes of SWCNTs–NH_2, SWCNTs–DoTAB, SWCNTs–SDS are similar and are confirmed from the SEM and TEM images displayed in Figure 10.7.

The poly disparity index (PDI) was measured for all the dispersed samples and found to be lower for SWCNTs-NH_2/DTAB and higher for SWCNTs-NH_2/SDS, compared to CNTs without any dispersion of surfactants. The PDI confirms the good dispersion of SWCNTs-NH_2/SDS. On the other hand, good dispersion depends on the electrostatic interactions developed between

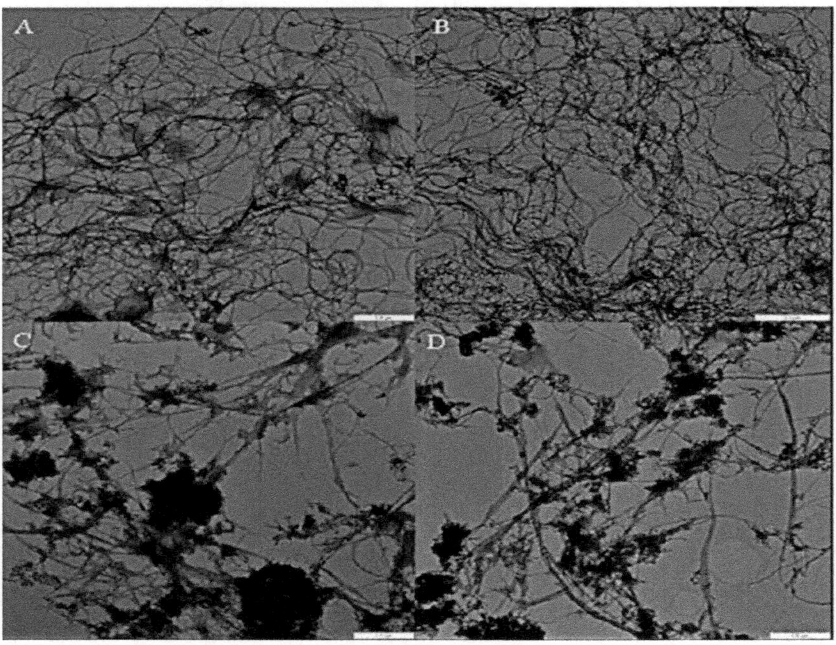

Figure 10.6 Transmission electron microscopic images of MWCNT/SDS (A), MWCNT/DoTAB (B), SWCNT/SDS (C) and SWCNT/DoTAB (D) with scale bars of 0.5 μm. The samples were prepared in a buffer solution Na_2HPO_4 0.01 M/NaCl 0.02 M, pH 9. Reproduced from ref. 24, https://doi.org/10.3390/ijms22020826, under the terms of the CC BY 4.0 License, http://creativecommons.org/licenses/by/4.0/.

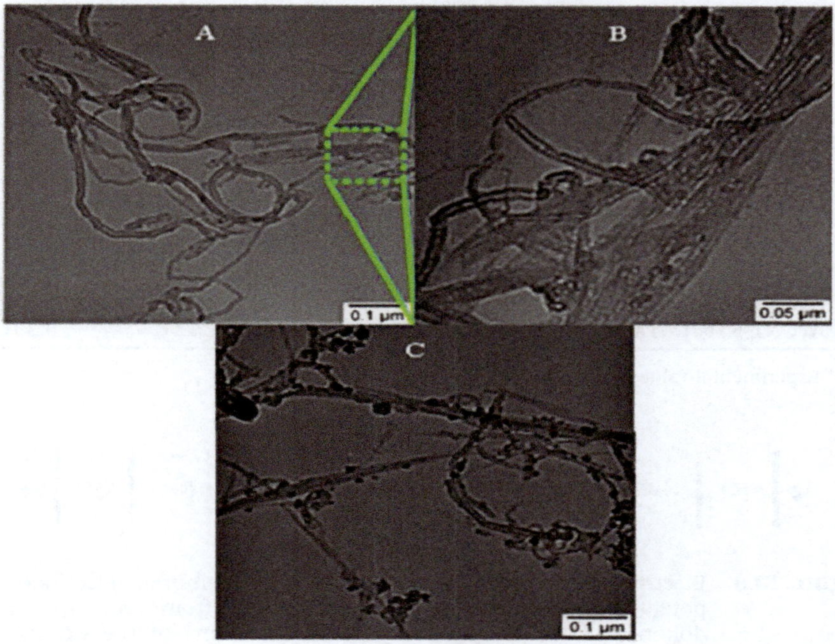

Figure 10.7 Transmission electron microscopic images of SWCNT-NH_2/SDS (A and B) and SWCNT-NH_2/DoTAB (C). Image B shows amplification of the area marked in green in image A. The samples were prepared in a buffer solution NaH_2PO_4 0.01 M/NaCl 0.02 M, pH 3. Reproduced from ref. 24, https://doi.org/10.3390/ijms22020826, under the terms of the CC BY 4.0 License, http://creativecommons.org/licenses/by/4.0/.

CNTs and surfactants and their degree of dissociation. TEM images of SWCNTs-DoTAB and SWCNTs-NH_2/DoTAB are displayed in Figure 10.6.

CNTs which are functionalized with DoTAB are better dispersing agents compared to non-functionalized CNTs, provided that the concentration of the dispersion agents remains the same. Table 10.1 summarizes the PDI values of the various dispersion SWCNTs with surfactants and their zeta potentials (ζ).

10.3.4 Potentiometric Measurements and Zeta Potentials

Ion-selective electrodes with different surfactants were analyzed by potentiometric estimation. Cationic ion-selective electrode membranes (HTAB, DoTAB, and OTAB) and anionic (SDS) surfactants were synthesized using the procedure reported in the previous literature.[31] The corresponding surfactants were placed in the buffer solution and immersed with ion-selective electrodes for about 2 h for conditioning. Figure 10.8 displays the cell representation of the various surfactants used in the potentiometric estimation. The concentrations of the surfactants used were OTAB 2×10^{-3} mol dm^{-3},

Table 10.1 Zeta-potential (ζ), size (d_H) and polydispersity index (PDI) values for different carbon nanotubes in the presence and absence of ionic surfactants. Reproduced from ref. 24, https://doi.org/10.3390/ijms22020826, under the terms of the CC BY 4.0 License, http://creativecommons.org/licenses/by/4.0/.

CNT/surfactant	ζ/mV[a]	d_H/nm	PDI
SWCNT/DoTAB	4.1 ± 0.2	1180 ± 30	0.455
SWCNT/SDS	-3.9 ± 0.3	1031 ± 25	0.556
MWCNT/DoTAB	1.7 ± 0.1	1483 ± 27	0.598
SWCNT/SDS	-4.5 ± 0.3	1194 ± 32	0.664
SWCNT-NH_2	13.3 ± 0.4	928 ± 15	0.470
SWCNT-NH_2/DoTAB	22.5 ± 0.2	920 ± 12	0.391

[a] Experimental values.

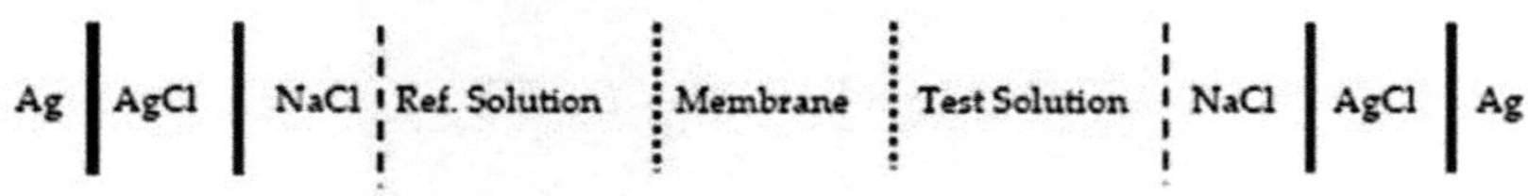

Figure 10.8 Electrochemical cell based on ion-selective membrane electrodes for potentiometric measurements. Reproduced from ref. 24, https://doi.org/10.3390/ijms22020826, under the terms of the CC BY 4.0 License, http://creativecommons.org/licenses/by/4.0/.

DoTAB 1×10^{-3} mol dm^{-3}, HTAB and SDS 1×10^{-3} mol dm^{-3}. Emfs were measured using an INA-116 ultra-low input bias current amplifier fitted with a Sallen key low pass filter. The response was recorded with a Keithley 2110 digital multi-meter connected to a PC having accuracy of ± 0.1 mV. The temperature was maintained constant using a thermostat placed in the solution. The calibration curve is obtained by plotting the concentrations of the free surfactant *vs.* the ion-selective membrane electrode using the Nernst eqn (10.1).

A Zetasizer Nano NS Malvern Instrument was used for the determination of zeta potentials (ζ) of the cationic and anionic surfactants dispersed with CNTs. The electrophoretic mobility of the particles was investigated using a Laser Doppler Velocimeter. Polycarbonate capillary wells (DTS1060) with a temperature of 298.0 ± 0.1 K were used in the experiments. The average of five successive experimentations was noted. SWCNTS-NH_2 measurements were carried out in the presence and absence of surfactants. The zeta potential graphs are displayed in Figure 10.9.

10.3.5 Influence of Membrane Potential

10.3.5.1 Influence of Composition of the Membrane

The selectivity and sensitivity of the ionosphere (DTA^+SD^-) depends not only the PVC membrane but also on the composition of the membrane ingredients, as reported in previous literature.[32,33] The initial step of the process involves testing various plasticizers, PVC and various ingredients on the ionosphere. The prepared membranes were studied against SDS concentration. Table 10.2 summarizes the lifetime study of the anionic surfactant-selective electrode. Membranes 1, 2, and 3 had similar compositions but differed in the ion pair content. Figure 10.9 shows the response of the content of the membrane electrodes 1, 2, and 3 *vs.* SDS concentration and

Table 10.2 Lifetime study of anionic surfactant-selective electrodes. Reproduced from ref. 25, https://doi.org/10.4061/2011/958647, under the terms of the CC BY 3.0 license, https://creativecommons.org/licenses/by/3.0/.

Membrane[a]	Slope (mV per decade)	Linear range ($mol\,L^{-1}$)	Detection limit ($mol\,L^{-1}$)	Correlation coefficient (r)
1	50.7 ± 1.2	7.5×10^6 to 2.5×10^3	4.6×10^6	0.9972 ± 0.0004
2	58.0 ± 0.9	5.0×10^6 to 2.5×10^3	2.9×10^6	0.9987 ± 0.0003
3	65.5 ± 2.4	7.5×10^6 to 2.5×10^3	5.9×10^6	0.9935 ± 0.0007

[a] Ingredients of membrane (wt%): (1) DOP (60%), PVC (34%), ion pair (1%), (2) DOP (60%), PVC (33%), ion pair (2%), (3) DOP (60%), PVC (32%), ion pair (3%).

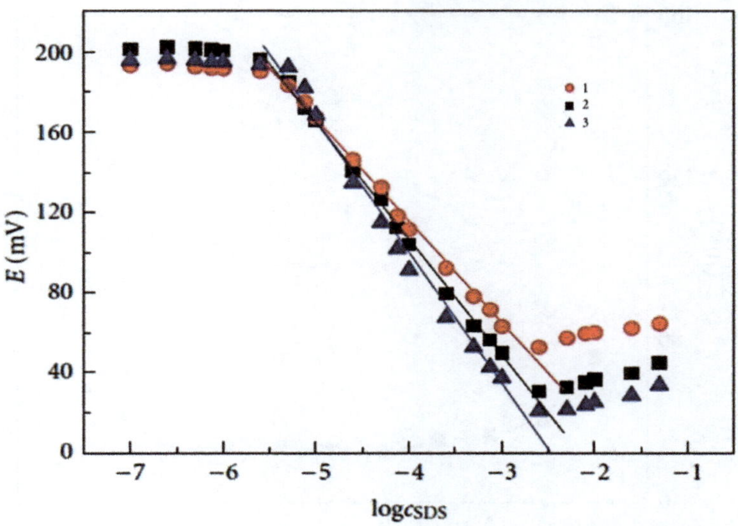

Figure 10.9 Potential response of the surfactant-selective electrode with different compositions of SDS. Ingredients (wt%): DOP (60%), PVC (34%), ion pair (1%), DOP (60%), PVC (33%), ion pair (2%), DOP (60%), PVC (32%), ion pair (3%). Reproduced from ref. 25, https://doi.org/10.4061/2011/958647, under the terms of the CC BY 3.0 license, https://creativecommons.org/licenses/by/3.0/.

summarizes the obtained results during the potentiometric estimation. Electrode 2 showed better performance in terms of wt% composition, 33% PVC, 60% DOP, and 2% ion pair, as compared to electrodes 1 and 3. The detection was at the lowest possible limit and slope value corresponding to the obtained theoretical value (59.2 mV per decade).

10.3.5.2 Electrode Response Behaviour

The emf developed by electrode 2 in SDS solution was investigated using the Nernst equation for single electrode potentials. Figure 10.10 displays the response characteristics of anionic surfactants *vs.* SDS concentration. Linear regions were used to calculate the slope and corresponding coefficients. The detection limit corresponds to the intersection of the linear graph by extrapolation of the straight line.[34] The response was found to be -58.9 ± 0.9 mV per decade with 5×10^{-6} mol L^{-1} and 2.5×10^{-3} mol L^{-1} SDS concentrations.

10.3.5.3 Response Time and pH

A series of SDS solutions were prepared with concentrations of 1.0×10^{-5} mol L^{-1} to 1.0×10^{-3} mol L^{-1} and the static response times were

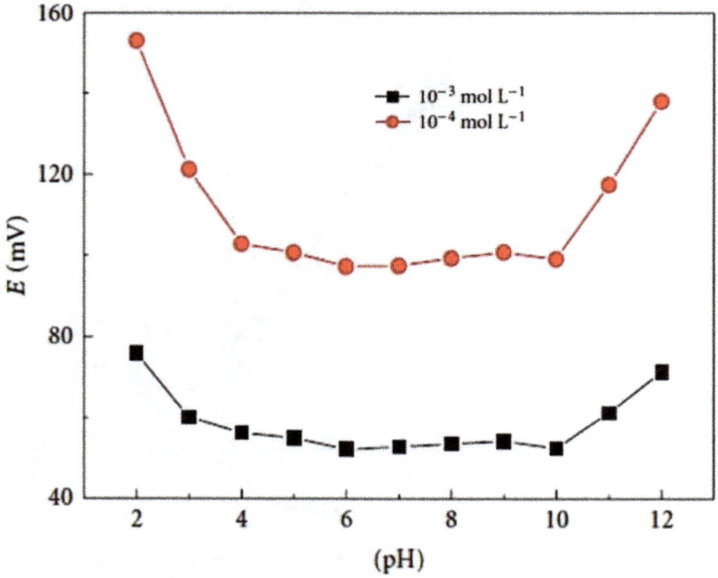

Figure 10.10 Effect of pH on the potentials of anionic surfactant selective electrode in 1×10^{-3} and 1×10^{-4} mol L^{-1} SDS solution. Reproduced from ref. 25, https://doi.org/10.4061/2011/958647, under the terms of the CC BY 3.0 license, https://creativecommons.org/licenses/by/3.0/.

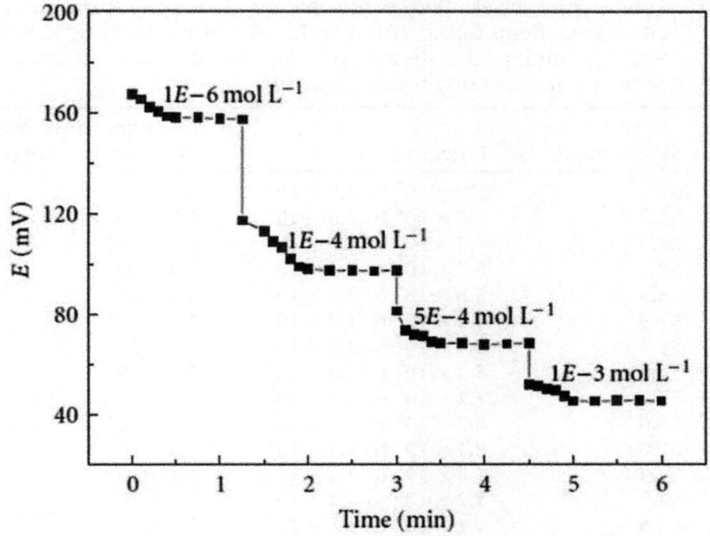

Figure 10.11 Response time of an anionic surfactant-selective electrode for step changes in concentration of SDS solution. Reproduced from ref. 25, https://doi.org/10.4061/2011/958647, under the terms of the CC BY 3.0 license, https://creativecommons.org/licenses/by/3.0/.

investigated. Response time is the time required by the electrode to achieve a steady state.[35,36] Figure 10.11 displays a response time of 40 s required by the electrode to achieve ± 1 mV of the final equilibrium. The standard deviation is calculated by recording for four different concentrations of the solution.

The electrode response with pH (for pH range 2–12) was studied with fixed concentrations of SDS (1.0×10^{-4} mol L^{-1} and 1.0×10^{-3} mol L^{-1}). For ion pairs of $CTA^{+}DS^{-}$, at low pH, DS^{-} combines with H^{+} to form DSH and the concentration of DS^{-} decreases. At higher pH values (pH > 10), the electrode potential increases due to the presence of an excess of OH^{-} ions in solution as a result of competition between OH^{-} and DS^{-} anions. As illustrated in Figure 10.11, in the range pH 4–10, the electrode potential does not remain the same and becomes constant. The pH range 4–10 was chosen for the study.[37]

10.3.5.4 Potentiometric Selectivity Coefficients (PSC) Studies and Lifetime of the Electrodes

Table 10.3 and Figure 10.12 summarize the electrode characteristics obtained at various time intervals. Good slope, linear concentration rage, and detection limit were observed for the electrode for up to 13 days; after 14 days there were slight drifts in the values for the slope, linear response, and detection limit. The electrode response in relation to SDS concentration

Table 10.3 Potentiometric selective coefficients for an anionic surfactant-selective electrode. Reproduced from ref. 25, https://doi.org/10.4061/2011/958647, under the terms of the CC BY 3.0 license, https://creativecommons.org/licenses/by/3.0/.

Period	Slope (mol L^{-1})	Linear range	Detection limit (weeks) (mV per decade) (mol L^{-1})
1	58.0	5.0×10^6 to 2.5×10^3	2.9×10^6
2	58.2	4.6×10^6 to 2.5×10^3	2.9×10^6
3	58.4	4.7×10^6 to 2.5×10^3	3.0×10^6
4	58.1	5.2×10^6 to 2.5×10^3	3.0×10^6
5	58.3	5.5×10^6 to 2.7×10^3	3.4×10^6
6	57.9	5.9×10^6 to 1.9×10^3	2.2×10^6
7	58.4	6.4×10^6 to 2.2×10^3	3.9×10^6
8	58.6	5.4×10^6 to 1.8×10^3	2.6×10^6
9	57.8	7.1×10^6 to 1.5×10^3	4.9×10^6
10	58.0	6.7×10^6 to 2.8×10^3	3.5×10^6
11	57.6	8.1×10^6 to 2.4×10^3	4.3×10^6
12	57.4	7.8×10^6 to 3.1×10^3	4.1×10^6
13	57.2	8.2×10^6 to 3.4×10^3	5.0×10^6
14	54.6	9.1×10^6 to 3.5×10^3	6.1×10^6
15	52.1	1.1×10^5 to 2.5×10^3	7.6×10^6
16	41.6	2.4×10^5 to 3.1×10^3	1.6×10^5

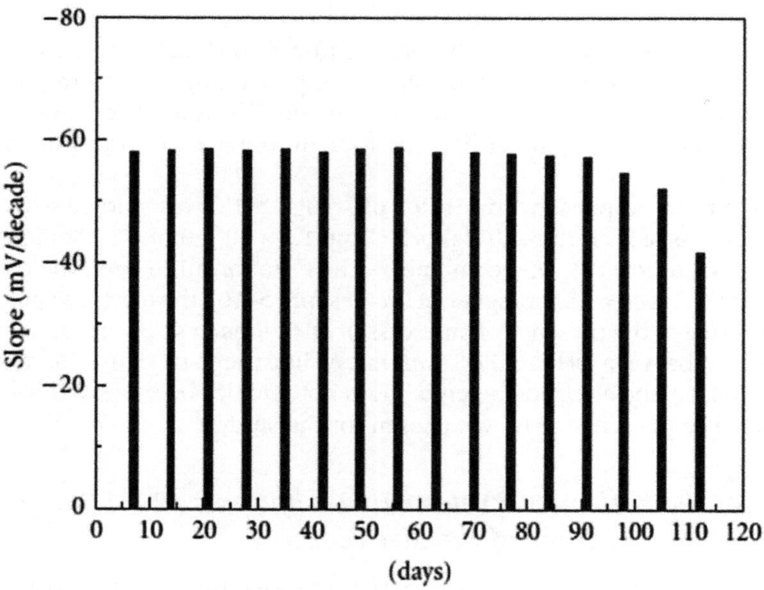

Figure 10.12 Plot of the slope of anionic surfactant selective electrode in the presence of SDS *versus* time (days). Reproduced from ref. 25, https://doi.org/10.4061/2011/958647, under the terms of the CC BY 3.0 license, https://creativecommons.org/licenses/by/3.0/.

Table 10.4 Results of determining the concentration of SDS by potentiometric titration using an anionic surfactant selective electrode compared with two-phase titration in sample solutions. Reproduced from ref. 25, https://doi.org/10.4061/2011/958647, under the terms of the CC BY 3.0 license, https://creativecommons.org/licenses/by/3.0/.

Entry	V_{SDS} (mL)	$V_{Hyamine}$ (mL)	CSDS (mol L^{-1})	$V_{Hyamine}$ (mL)	CSDS (mol L^{-1})
1	25.0	2.53	4.09×10^{4}	2.51	4.06×10^{4}
2	25.0	2.54	4.11×10^{4}	2.52	4.08×10^{4}
3	25.0	2.54	4.11×10^{4}	2.53	4.09×10^{4}
Average	25.0	2.54	4.11×10^{4}	2.51	4.08×10^{4}
RSD^{a} (%)			1.41	1.58	

^{a}Relative standard deviation of potentiometric method for triplicate measurements.

was studied. Figure 10.12 displays the selective response to SDS and poor change in the potentials for F^-, Cl^-, Br^-, NO_3^-, SO_4^{2-}, *etc.* It can be seen that the electrode response was less for these anions compared to SDS. Determination of potentiometric selectivity coefficients gives information regarding the electrode selectivity towards SDS. By considering SDS as the principle anion, with 2.5×10^{-2} mol L^{-1} concentration, using the interfering anions and from eqn (10.2).

$$K_{ij}^{\text{plot}} = \frac{a_{DS^-}}{(ax^-)^{\frac{Z_{DS^-}}{Z_{X^-}}}} \tag{10.2}$$

where a_{DS^-} is the activity of the primary ion of DS, a_{X^-} is the activity of the corresponding interfering ion, and Z_{DS^-} and Z_{X^-} are the corresponding charge of the primary ion and the interfering ion, respectively.

Table 10.4 summarizes the potentiometric selectivity coefficients. From Table 10.4, the logarithm of the potentiometric selectivity coefficients determined for the electrode in the present study were lower than 3.0 which indicates that most of the anions significantly disturbed the detection of SDS in the real samples.

10.3.5.5 Analytical Application Studies

The present study envisaged that the electrode would not only be useful for the potentiometric determination of SDS but could also be used in the presence of precipitating agents like hyamine. The accurate concentration of SDS determined by using hyamine (precipitating agent) in aqueous solution. In the first instance, the concentration of SDS in water was determined using a potentiometer in conjunction with an SCE. The obtained results were compared with the results of the two-phase titration method.[38,39] Figure 10.13 displays a sigmoid-type curve obtained during the titration of SDS. The titration volume was obtained by the reflection point in the

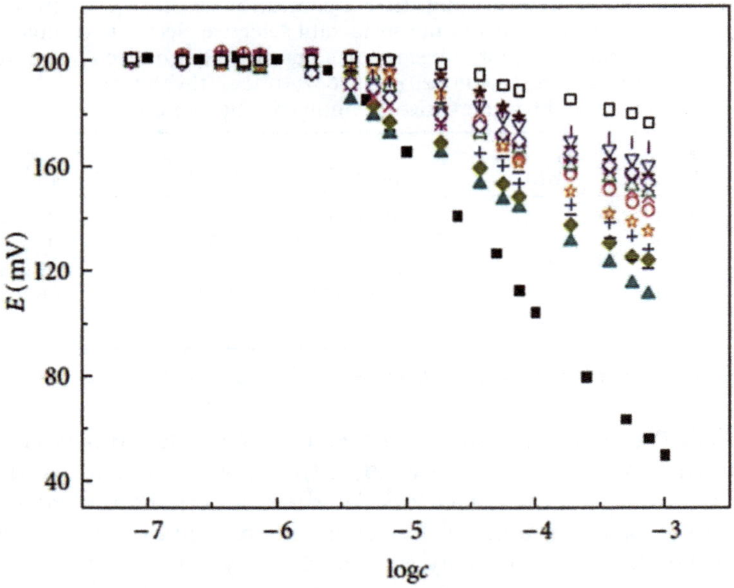

Figure 10.13 Response of an anionic surfactant-selective electrode in the presence of certain anions: SDS; F; (+) Cl; Br; I; NO_3; SO_4^{2-}; $PO4^{3-}SO_3^{2-}$; CO_3^{2-}; HCO_3; CH_3COO^-, $C_2O_4^{2-}$, $C_4H_4O_6^{2-}$; $C_6H_5O_7^{3-}$. Reproduced from ref. 25, https://doi.org/10.4061/2011/958647, under the terms of the CC BY 3.0 license, https://creativecommons.org/licenses/by/3.0/.

Table 10.5 Comparison of the results obtained by the potentiometric and two-phase titration of commercial detergents. Reproduced from ref. 25, https://doi.org/10.4061/2011/958647, under the terms of the CC BY 3.0 license, https://creativecommons.org/licenses/by/3.0/.

Sample SDS (wt%)	Potentiometric titration	Two-phase titration
1[a]	14.37	14.23
2[a]	14.34	14.20
3[a]	14.39	14.26
RSD[b] (%)	2.05	2.45

[a] Number 1,2,3: washing liquid (SDS, glycerine, sodium chloride, diethanol amide, and additives).
[b] Relative standard deviation of potentiometric method for triplicate measurements.

titration curve. Table 10.5 summarizes the concentrations of the surfactants with SDS in the two-phase titration experiments. The standard deviations for potentiometric titration and two-phase titration were found to be 1.6%. In a typical procedure, commercial detergents (15–30 mg) were dissolved in 2 mL of methanol and diluted to 100 mL; 20 mL was used for the estimation

Table 10.6 Comparison of the results of slope, linear range, and detection limit of various ionophores. Reproduced from ref. 25, https://doi.org/10.4061/2011/958647, under the terms of the CC BY 3.0 license, https://creativecommons.org/licenses/by/3.0/.

Ionophore	Slope (mV per decade)	Linear range ($mol\,L^{-1}$)	Detection limit
New cyclic aza-oxa-cycloalkane[40]	57.7 ± 0.2	3.3×10^{6} to 6.7×10^{3}	2.9×10^{6}
Cyclam derivative[41]	60.0 ± 0.9	7.9×10^{6} to 2.0×10^{3}	4.0×10^{6}
Single-walled carbon nanotubes[42]	59.5	9.0×10^{5} to 5.0×10^{3}	5.2×10^{6}
CTAD Sion pair[43]	58.0 ± 0.9	5.0×10^{6} to 2.5×10^{3}	2.9×10^{6}

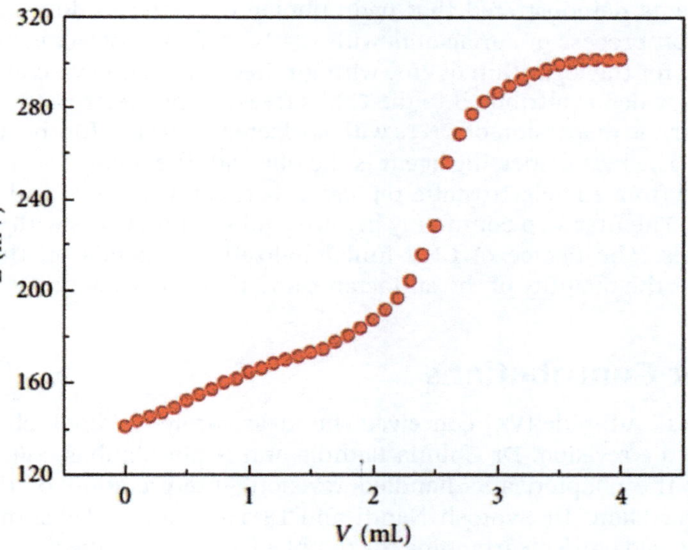

Figure 10.14 Potentiometric titration curve of 25.0 mL tested SDS solution with 4.043×10^{-3} $mol\,L^{-1}$ hyamine as titrant in water, using the electrode as an indicator electrode. Reproduced from ref. 25, https://doi.org/10.4061/2011/958647, under the terms of the CC BY 3.0 license, https://creativecommons.org/licenses/by/3.0/.

of SDS using ion-selective electrodes. The obtained results from the potentiometric titrations were compared with those obtained from the two-phase titration process. Table 10.5 summarizes potentiometric titrations and two-phase titration results of commercial detergents. Table 10.6 summarizes a comparison of similar reports in the literature (Figure 10.14).

10.4 Conclusions

In summary, the developed membrane cyclic aza-oxa cycloalkane ionosphere has the widest linear applications with the lowest detection limit. The ion-selective membrane with the CTA^+DS^- ion pair has the widest pH range with the lowest detection limit. The ion-selective electrode using the CTA^+DS^- ion pair using surfactant possesses easy preparation, and a fast and clean method of analysis with respect to the surfactants. The ion-selective electrode displayed a Nernstian slope of -8 ± 0.9 mV per decade in a 5.0×10^{-6} to 2.5×10^{-3} mol L^{-1} concentration range and a limit of detection of 2.9×10^{-6} mol L^{-1}. The electrode also showed clear responses for te anionic surfactants also applicable in the pH range of 4–10 with a response time of 40 s $\pm$ 5 s. Common anions do not interfere in the SDS titration, which can be confirmed from the potentiometric selectivity coefficients. On the other hand, with the help of ion-selective electrodes using potentiometric techniques, the interaction of commercial ionic surfactants with CNTs has been investigated. The experiments demonstrated that hydrophobic character is dominant in the adsorption process of surfactants with CNTs. A linear correlation curve was obtained for the logarithm of cmc with ion-free surfactants *vs.* concentration of CNTs, which is attributed to the Gibbs free energy. Electrostatic forces are useful for the dispersion of CNTs with surfactants but not for the adsorption process. The best dispersing agent is the one with the highest surface charge resulting from the electrostatic repulsion between the CNTs and ionic surfactants. The first step commonly involves unbundling CNTs with surfactant complexes. The degree of CNT functionalization depends on the concentration or the quantity of the surfactants and their dispersion.

Author Contributions

Dr Vinayak Adimule (VA) conceived the ideas, wrote the book chapter, and handled the revision. Dr Chinna Bathula and Nidhi Manhas contributed to drafting the chapter, and handled revision, interpretation of the results and supervision. Dr Santosh Nandi and Dr Parashuram Lakshminarayana were involved partly in arranging the template and grammatical and language corrections.

Conflicts of Interest

All authors declare that they do not have any conflict of interest.

Data Availability

The datasets generated during and/or analyzed during the current study are available from the corresponding author on reasonable request.

Funding

All authors declare that they have not received any funding from any source or organization.

Acknowledgements

All the authors are thankful to Nitte Meenakshi Institute of Technology, Bangalore, Karnataka, India and KLE Technological University, Dr MS Sheshgiri College of Engineering and Technology, Belagavi for supporting and encouraging for this work.

References

1. S. O. Badmus, H. K. Amusa and T. A. Oyehan, *Environ. Sci. Pollut. Res.*, 2021, **28**, 62085.
2. J. H. Yoon, Y. G. Shin, M. B. Kirkham, S. S. Jeong, J. G. Lee, H. S. Kim and J. E. Yang, *Toxics*, 2022, **10**, 162.
3. J. Alonso, J. Barb, J. Bartroli, J. Siinchez and M. de Valle, *Anal. Chim. Acta*, 1995, **308**, 115.
4. F. J. Ostos, J. A. Lebron, M. L. Moya, E. Bernal, A. Flores, C. Lepori, A. Maestre, F. Sanchez, P. Lopez-Cornejo and M. Lopez-Lopez, *Int. J. Mol. Sci.*, 2021, **22**, 826.
5. M. Novell, M. Parrilla, G. A. Crespo, F. X. Rius and F. J. Andrade, *Anal. Chem.*, 2012, **84**, 4695.
6. V. Adimule, S. Batakurki, B. C. Yallur, C. Bathula and L. Parashuram, *J. Mater. Sci.: Mater. Electron.*, 2022, **33**, 18699.
7. B. C. Yallur, V. Adimule, M. S. Raghu, F. A. Alharthi, B. Jeon and L. Parashuram, *Surf. Interfaces*, 2023, **36**, 102587.
8. M. Najafi, L. Maleki and A. A. Rafati, *J. Mol. Liq.*, 2011, **159**, 226.
9. M. J. Segui, J. L. Sabater, R. M. Manez, T. Pardo, F. Sancenom and J. Soto, *Anal. Chim. Acta*, 2004, **525**, 83.
10. (a) S. Rajendrachari, N. Basavegowda, V. Adimule, B. Avar, P. Somu, S. K. RM and K. Baek, *Biosensors*, 2022, **12**, 1173; (b) K. A. Kamath, J. G. Manjunatha, T. Girish, M. Sillanpaa, A. M. Tighezza and M. D. Albaqami, *Inorg. Chem. Commun.*, 2022, **143**, 109811.
11. G. Tigari, J. G. Manjunatha, E. D Souza, C. Raril, N. Hareesha and M. M. Charithra, *Monatshefte Chem.-Chem. Monthly*, 2022, **153**, 311.
12. S. Rajendrachari, V. Adimule, M. Gulen, F. Khosravi and K. K. Somashekharappa, *Materials*, 2022, **15**, 7591.
13. R. Shashanka, G. K. Jayaprakash, P. B. G. Mohan Kumar Kumar and B. E. Kumara Swamy, *Mater. Res. Innovations*, 2022, **26**, 229.
14. B. C. Yallur, S. S. Nandi and V. Adimule, *Adv. Mater. Res.*, 2022, **1173**, 59.
15. P. A. Pushpanjali, J. G. Manjunatha, N. Hareesha, B. M. Amrutha, C. Raril, Z. A. Alothman, A. M. Alanazi and A. Pandith, *Mater. Chem. Phys.*, 2022, **276**, 125392.
16. N. B. Pradeep, M. M. Rajath Hegde, S. Rajendrachari and A. O. Surendranathan, *Powder Technol.*, 2022, **408**, 117715.
17. P. A. Pushpanjali, J. G. Manjunatha, B. M. Amrutha and N. Hareesha, *Mater. Res. Innovations*, 2021, **25**, 412.

18. V. Adimule, S. Batakurki, B. C. Yallur and R. Keri, *J. Mater. Res.*, 2022, **37**, 2396.
19. N. Hareesha, J. G. Manjunatha, B. M. Amrutha, P. A. Pushpanjali, M. M. Charithra and N. P. Subbaiah, *J. Electron. Mater.*, 2021, **50**, 1230.
20. M. Lopez-López, E. Bernal, M. L. Moya, F. Sanchez and P. Lopez-Cornejo, *Electrochem. Commun.*, 2016, **67**, 31.
21. (a) M. Challa, S. Chinnam, A. M. Rajanna, A. Nandagudi, B. C. Yallur and V. Adimule, *Heliyon*, 2023, **9**, e13223; (b) V. Adimule, S. Batakurki, B. C. Yallur and R. Keri, *J. Mater. Res.*, 2022, **15**, 2396.
22. (a) V. Adimule, S. Batakurki, V. S. Bhat, B. C. Yallur, G. Hegde and C. Bathula, *J. Energy Storage*, 2023, **57**, 106318; (b) V. Adimule, S. S. Nandi, B. C. Yallur, D. Bhowmik and A. H. Jagadeesha, *Mater. Today Chem.*, 2021, **20**, 100438.
23. (a) V. Adimule, B. C. Yallur, R. Keri, C. Bathula and S. Batakurki, *Electron. Mater. Lett.*, 2022, 1; (b) V. Adimule, V. S. Bhat, B. C. Yallur, A. H. J. Gowda, P. De Padova, G. Hegde and A. Toghan, *Nanomater. Nanotechnol.*, 2022, **12**, 18479.
24. (a) V. Adimule, S. Rajendrachari, R. Mahale, S. Batakurki, B. C. Yallur, S. Nandi and G. Bagihalli, *Silicon*, 2022, 1; (b) F. J. Ostos, J. A. Lebrón, M. L. Moyá, E. Bernal, A. Flores, C. Lépori, A. Maestre, F. Sánchez, P. López-Cornejo and M. López-López, *Int. J. Mol. Sci.*, 2021, **22**(2), 826.
25. J. Wang, Z. Du, W. Wang and W. Xue, *Int. J. Electrochem.*, 2011, **2011**, 958647.
26. M. Novell, M. Parrilla, G. A. Crespo, F. Xavier Rius and F. J. Andrade, *Anal. Chem.*, 2012, **84**, 4695.
27. J. Wang, Z. Du, W. Wang and W. Xue, *Int. J. Electrochem.*, 2011, **7**, 958647.
28. G. G. Mohameda, T. A. Ali, M. F. El-Shahat, A. M. Al-Sabagh, M. A. Migahed and E. Khaled, *Anal. Chim. Acta*, 2010, **673**, 79.
29. J. Alonso, J. Barb, J. Bartroli, J. Siinchez and M. de1 Valle, *Anal. Chim. Acta*, 1995, **308**, 115.
30. (a) M. Lopez-Lopez, E. Bernal, M. L. Moya, F. Sanchez and P. López-Cornejo, *Electrochem. Commun.*, 2016, **67**, 31; (b) G. G. Mohameda, T. A. Ali, M. F. El-Shahat, A. M. Al-Sabagh, M. A. Migahed and E. Khaled, *Anal. Chim. Acta*, 2010, **673**, 79; (c) G. Tigari, J. G. Manjunatha, E. S. D'Souza and N. Sreeharsha, *ChemistrySelect*, 2021, **6**, 2700.
31. (a) J. Wang, Z. Du, W. Wang and W. Xue, *Int. J. Electrochem.*, 2011, **958647**, 7; (b) N. Hareesha, J. G. G. Manjunatha, C. Raril and G. Tigari, *Adv. Pharm. Bull.*, 2019, **9**, 132.
32. G. K. Jayaprakash, B. E. Kumara Swamy, S. Rajendrachari, S. C. Sharma and R. Flores-Moreno, *J. Mol. Liq.*, 2021, **334**, 116348.
33. R. Shashanka and D. Chaira, *Tribol. Trans.*, 2017, **60**, 324.
34. J. G. Manjunatha, B. E. Kumara Swamy, G. P. Mamatha, C. Raril, L. Nanjunda Swamy and S. Fattepur, *Mater. Today: Proc.*, 2018, **5**, 22368.
35. V. Adimule, B. C. Yallur and A. Gowda, *Anal. Bioanal. Electrochem.*, 2022, **14**, 1.

36. R. S. Keri, V. Adimule, P. Kendrekar and B. S. Sasidhar, *Top. Catal.*, 2022, 1.
37. V. Adimule, S. S. Nandi and and A. H. Jagadeesha Gowda, *Proceedings of the 3rd International Conference on Advanced Technologies for Societal Applications*, 2021, vol. 2, p. 55.
38. B. C. Yallur, V. Adimule, A. Suryavanshi and S. Nandi, *Macromol. Symp.*, 2020, **392**, 2000001.
39. V. Adimule, G. B. Bagihalli and S. N. Unki, *Adv. Mater. Res*, 2022, **1173**, 3.
40. M. J. SeguI, J. Lizondo-Sabater, R. Martínez-Manez, T. Pardo, F. Sancenon and J. Soto, *Anal. Chim. Acta*, 2004, **525**, 83.
41. M. P. Lima, D. Esteban-Gómez, R. Delgado, C. Platas-Iglesias and R. Tripier, *Inorg. Chem.*, 2012, **51**, 6916.
42. P. Deria, J. Olivier, J. Park and M. J. Therien, *J. Am. Chem. Soc.*, 2014, **136**, 14193.
43. Z. Li and M. J. Rosen, *Anal. Chem.*, 1981, **53**, 1516.

CHAPTER 11

Advantages, Limitations, Future Expectations, and Commercialization Challenges of Surfactant Sensors

P. KARPAGAVINAYAGAM, S. ALWIN DAVID AND C. VEDHI*

V.O. Chidambaram College, Thoothukudi – 628 008, Tamilnadu, India
*Email: cvedhi23@gmail.com

11.1 Introduction

A sensor is an instrument that recognizes and reacts to input from the external surroundings *viz.*, illumination, temperature, vibration, humidity, strain, or any amount of other physical parameters can be utilized as inputs. The development of sensing technology has also resulted in low-cost, small, multifunctional sensing properties. These devices are composed of a large number of arbitrary tiny sensor nodes dispersed across a geographic region, which interact with one another to form a sensing network. By collecting, processing, analysing, and distributing data, a sensor network can provide data access at any time and from any location. As a result, the system keenly contributes to the creation of a smart environment (see Figure 11.1).

Sensors offer a wide range of benefits, opportunities and applications, including enhanced sensitivity, selectivity, stability, compatibility, signal amplification, and miniaturization. Continued research and development in this field are likely to yield further advancements and novel applications in sensing technologies. Surfactant-stabilized substances, perspectives, as well

Detection Science Series No. 23
Surfactant-based Sensors in Chemical and Biochemical Detection
Edited by J. G. Manjunatha

Published by the Royal Society of Chemistry, www.rsc.org

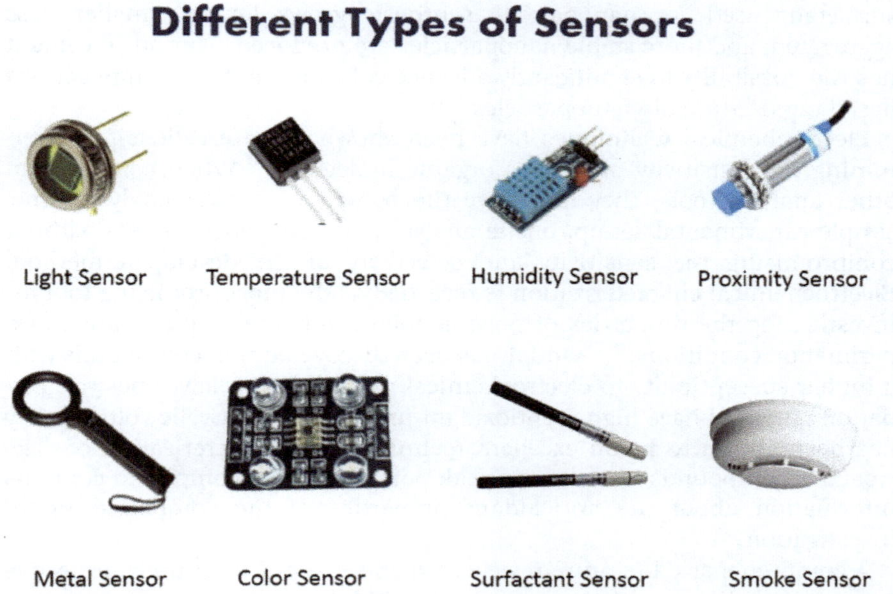

Figure 11.1 Types of sensor.

as poses a potential are widely demonstrated and presented in detail for their use in sensing applications. Based on the number of carbon atoms present and the complexity of the emulsifier, they is split into two categories: typical and advanced surfactants. Surfactants have long been recognized as one of the most adaptable process chemicals in industrial biotechnology. Surfactants as organized arrangements are widely used in daily life, chemical production, health sciences, and materials science, as well as in other important applications such as nanoreactors, drug carriers, and detector fabrication. In this regard, biosurfactants may be the best potential compounds. Different properties can be used to identify and sense a wide range of analytes, including heavy metal ions, alkaline earth metals, transition metals, organic molecules, microorganisms, and oligonucleotides, among others. Surfactants are widely used to lower surface and interfacial tensions among a number of phases owing to their structural uniqueness.[1–4] Their proclivity to form self-assembled structures in solution can also result in the formation of micelles with diameters of a few micrometres, ranging between nanometers and microns.[5] The amphiphilic nature of surfactants makes them suitable for use in a wide range of industrial goods, including medicinal products, corrosion inhibitors for steel protection, and other materials. Chemicals include corrosive metals,[6,7] detergents,[8,9] de-emulsifiers,[10,11] wetting agents,[12,13] oil recovery enhancers,[14,15] pour-point depressants, pharmaceutical formulations, and materials used in drug delivery. Surfactants are crucial in improving dispersion and trying to control the size of as-synthesized nanoparticles. More specifically, as the tail length of the surfactant used for nanoparticle synthesis grows longer, smaller, less aggregated, and more stable nanoparticles are produced. This advancement has the possibility to significantly advance colorimetric sensor applications based on size-tunable nanoparticles.[16,17]

Electrochemical techniques have been shown to be excellent for determining the sensitivity of various organic molecules.[18] When compared to other analysis tools, they also have the benefit of shorter analysis time, simple experimental set-up, onsite analysis, and cost effectiveness without compromising the sensitivity and selectivity of the developed method. Electrochemical characterization is regarded as the most promising tool for investigating the properties of polyphenolic compounds under various experimental conditions,[19,20] and it has been discovered that compounds with a higher susceptibility to electrochemical oxidation or a lower positive oxidation potential have higher antioxidant properties.[21–24] Cyclic voltammetry has been shown to be an excellent technique for characterization of electroactive compounds because the peak potential of the compound contains information about the antioxidant properties of the compound under investigation.

A comprehensive literature review of such sensors within their respective categories is discussed below as separate subheadings. An overview of electrochemical surfactant sensors is provided, with a focus on scientific publications since the 1990s. The significance of surfactants in modern biotechnology is emphasized. Amperometric surfactant sensors are real biosensors that use microorganisms or residing cell lines, as opposed to potentiometric, amperometric, conductometric, and impedimetric sensors.[25–30]

Contemporary biotechnology's buzzwords are biosurfactants, biotransformation, microbial degradation, and bioavailability. Surfactants are commonly used in advanced textile technologies and biotech.[31,32] The use of enzymes in disinfectants (primarily protease and amylase) necessitates bioengineering expertise to address issues such as enzymatic activity, as well as separation, enzyme oxidative stability and action preservation, thermostability, pH stability, and so on.

11.2 Advantages and Challenges of Sensors

The utilization of surfactants as a stabilization medium for nanomaterials is one of the oldest approaches used to identify the aggregation problem. In the peer-reviewed literature, the word surfactant is used broadly to refer to a wide range of surface-active materials. Surfactants are defined as any substance that can reduce the surface tension of a material by direct interaction at the interface due to the presence of both hydrophobic and hydrophilic components in its structure.[33]

Voltammetric sensors are simple, potent, as well as inexpensive devices with simple principles that have been broadly used in electrochemical studies to investigate the properties of electroactive species. Surfactants are frequently used in the fabrication of voltammetric sensing devices to improve selectivity and sensitivity, which has a serious influence on electrostatic functionality in terms of dissolving organic compounds, creating a specific orientation of molecules on the electrode surface, and enhancing electron transfer reactions.[34]

Potentiometric sensors determine analyte concentration primarily by assessing the variation of the potential difference between working and reference electrodes at various analyte concentrations. Potentiometric surfactant sensors are primarily used as even though some straightforward potentiometric titration end-point sensors. The most well-liked, this category includes coated wire and liquid sensors, ISFET (ion-selective field effect transistor) and membrane sensors. The sensors are appropriate for determining ionic (cationic and anionic) and non-ionic surfactants. They can be used intermittently in batch mode or in flow injection analysis (FIA) mode continuously. The line between coated wire sensors and liquid membrane surfactant sensors is blurred. A transducer (or sensor) that converts the activity of a specific ion dissolved in a solution into an electrical potential is known as an ion-selective electrode (ISE). According to the Nernst equation, voltage is theoretically dependent on the logarithm of ionic activity. Ion-selective electrodes are used in analytical chemistry and biochemical/biophysical research to measure the concentration of ions in an aqueous solution. Both types rely on ISE membranes. Because these membranes are usually solid for practical reasons, the term "liquid membrane" can sometimes be misrepresentative. Nanomaterials (NMs) can be used for identification and detection of a variety of analytes, including heavy metal ions, alkaline earth metals, transition metals, organic molecules, microorganisms and oli. A detailed examination of the literature on such sensors within their respective categories has been done.[35,36]

In past few decades, considerable educational literature has been recorded on the synthesis of surfactant-stabilized gold (Au) nanomaterials (NMs) using surfactants. One of the applications is the use of surfactant-stabilized Au NMs as sensors that detect the presence of pollutants. Modulation of the morphologies of the NMs at the nanoscale level is required for the utilization of exciting and peculiar properties of NMs, as applications connected with these NMs cannot be effectively developed if long-term stability of the synthesized size and shape of the NMs is not achieved.[37] Surfactants continue to be the best shape-directing agent for the shape or size modulation of the materials used for this purpose.

A colorimetric sensor was developed to detect and assessment nitroaromatic compounds, specifically TNT and tetryl, by using modified metal nanoparticles (MNPs) stabilized by surfactant and functionalized with diethyldithiocarbamate (DDTC). The modification agents for MNPs were the cationic surfactants cetyl trimethyl ammonium bromide (CTAB) and DDTC incorporated into the MNP synthesis during chemical reduction with $NaBH_4$.

To increase the sensitivity and lower the detection limits of voltammetric methods, pulse methods have been introduced. Studies have proven that DPV and SWV are the most efficient in assessing trace amounts of electrocatalytic compounds in biofluids and medical products, with significant

Table 11.1 Challenging sensors.[a]

Technique	Surfactant	Working electrode	Samples	Ref.
CV	CTAB	CTAB/CPE	PVC water bottle	38
CV	CTAB	CTAB/CPE	ANT	41
CV	SDS	PGE	Urine	42
CV	Tx 100	TX-100/CPE	Serum	43
LSV	CTAB	CTAB/ABPE	Tablet	44
DPV	CTAB	GCE	Tablet serum	45
DPV	CTAB	GCE	Urine serum	46
SWV	CTAB	CB/GCE	Lake water	47
SWV	SDS	GCE	Serum	48
SWAdSV	SDS	PGE	Plasma	49
SWCAdSV, DPCAdSV	CTAB	HMDE	Wastewater	50

[a] Cyclic voltammetry (CV), linear sweep voltammetry (LSV), differential pulse voltammetry (DPV), squarewave voltammetry (SWV), cetyltrimethylammonium bromide (CTAB), sodium dodecyl sulfate (SDS), Triton X-100 (Tx100), carbon-paste electrode (CPE), glassy carbon electrodes (GCE), pencil graphite electrode (PGE), hanging mercury drop electrode (HMDE).

advantages such as ease of use and low equipment costs. Using the technique of cyclic voltammetry (CV), Lu *et al.* created an electrochemical sensor based on CTAB to estimate nonylphenol (NP) in a carbon paste electrode (CPE). In another study by Khashaba *et al.*, sodium dodecyl sulfate (SDS) surfactant was added to an electrolyte solution with rabeprazole sodium (RAB sodium), and the impact on the voltammetric response was assessed using a pencil graphite electrode (PGE).[38] It was concluded that 45 μM SDS could enhance the oxidation peak current. Employing CV for measurement, the linear range was found to be within the range 0.5–250 μM (limit of detection: 0.18 μM).

A whole-cell sensing element was used to create an innovative method for determining anionic surfactants. Many biosensors have been created to track essential ecological particles, such as biological oxygen demand (BOD), ammonia, and pesticides.[39]

Francisco *et al.*, described the first use of films of surfactant-stabilized Au and AuAg nanoparticles (NPs) for chemiresistive sensing of VOCs. The particles were chemically synthesized in the presence of tetraoctylammonium bromide (TOABr), drop-cast deposited between a pair of electrodes, and then tested for chemiresistive responses to different VOCs[40] (see Table 11.1).

11.3 Applications of Surfactant Sensors

Fogliatto *et al.* documented the synthesis of a cetyl trimethylammonium bromide-stabilized hanging mercury drop electrode for the detection of genistein with an LOD of 34.3 nM.[51] Jain *et al.* have reported the synthesis of a Tween 20-stabilized hanging mercury drop electrode for the detection of entacapone with an LOD of 0.42 nM.[52] Sanghavi and co-workers have synthesized a triton X-100-stabilized multiwall carbon nanotube paste electrode and used it as sensor for aspirin and caffeine with LODs of 0.084 μM and 0.088 μM respectively.[53]

Sataraddi *et al.* have investigated the synthesis of cetyl trimethylammonium bromide-stabilized glassy carbon electrode for 5-fluorouracil with an LOD of 20.13 nM.[54] In another report, sodium dodecyl sulfate-stabilized boron-doped diamond with an LOD of 2.86 nM was prepared by Yardım and co-workers and used for the detection of BaP.[55] Jain *et al.* documented the synthesis of a Tween-20-stabilized hanging mercury drop electrode for the detection of drotaverine hydrochloride with an LOD of 4.14 nM.[56]

Deng *et al.* have reported the synthesis of cetyl trimethylammonium bromide-stabilized etrabromobisphenol for the detection of ascorbic acid and rutin with LODs of 1.0 μM and 4.0 nM, respectively.[57] Dar and co-workers synthesized a cetyl trimethylammonium bromide-stabilized hanging mercury drop electrode and used it as sensor for quinine with an LOD of 0.40 μM.[58] Duarte *et al.* have investigated the synthesis of a cetylpyridinium bromide-stabilized carbon nanotube paste electrode for acetaminophen and ascorbic acid with respective LODs of 7.1 μM and 2.1 μM.[59]

In another report, cetyl trimethylammonium bromide-stabilized graphene nano-sheet paste with an LOD of 0.6 μM was prepared by Liu and co-workers and used for the detection of dopamine.[60] Lu *et al.* documented the synthesis of a cetyl trimethylammonium bromide-stabilized carbon paste electrode for the detection of nonylphenol with an LOD of 0.01 μM.[61] Shankar *et al.* have reported the synthesis of a dodecyl benzene sulfate-stablized carbon paste electrode for the detection of dopamine with an LOD of 0.01 μM.[62]

Levent and co-workers have synthesized a cetyl trimethylammonium bromide-stabilized glassy carbon electrode and used it as a sensor for testosterone with an LOD of 1.18 nM.[63] Ziyatdinova *et al.* have investigated the synthesis of cetylpyridinium bromide/single wall carbon nanotube-COOH/ for the detection of morin with an LOD of 28.9 nM.[64] In another report, a cetyl trimethylammonium bromide-stabilized hanging mercury drop electrode with an LOD of 1.98 μM has been prepared by Rather and co-workers and used for the detection of cefitizoxime.[65]

Materon *et al.* documented the synthesis of a multi-wall carbon nanotube-COOH/sodium dodecyl sulfate/screenprinted carbon electrode for the detection of cisplatin with an LOD of 4.6 μM.[66] Gowda *et al.* reported the synthesis of a multi-wall carbon nanotube/cetyl trimethylammonium bromide/glassy carbon electrode for the detection of 4-aminoantipyrine with an LOD of 0.16 nM.[67] Fekry and co-workers have synthesized a sodium dodecyl sulfate-stabilized Ce/carbon paste electrode and used it as a sensor for nicotine with an LOD of 94.3 nM.[68]

Temerk *et al.* have investigated the synthesis of a cetyl trimethylammonium bromide-stabilized ZnO/carbon paste electrode for nilutamide with an LOD of 3.21 nM.[69] In another report, a cetyl trimethylammonium bromide-stabilized glassy carbon electrode with an LOD of 0.11 μM has been prepared by Nayak and co-workers and used for the detection of thiosalicylic acid.[70]

Zhang *et al.* documented the synthesis of a didodecyldimethylammonium bromide/expanded graphite paste electrode for the detection of bisphenol with an LOD of 7.1 nM.[71]

Khashaba *et al.* have reported the synthesis of a sodium dodecyl sulfate-stabilized pyrolytic graphite electrode for the detection of rabeprazole sodium with an LOD of 0.18 μM.[72] Deroco and co-workers have synthesized a cetyl trimethylammonium bromide-stabilized carbon black/glassy carbon electrode and used it as a sensor for mesotrione with an LOD of 0.02 μM.[73] Farag *et al.* have investigated the synthesis of a sodium dodecyl sulfate/carbon paste electrode for the detection of ethylvanillin and methylvanillin with LODs of 0.02 μM and 0.02 μM, respectively.[74]

In another report, sodium dodecyl sulfate-stabilized boron-doped diamond with an LOD of 0.01 μM has been prepared by Dönmez and co-workers and used for the detection of enrofloxacin.[75] Ozcelikay *et al.* documented the synthesis of a benzalkonium chloride-stabilized Ag-benzalkonium chloride/glassy carbon electrode for the detection of tenofovir with an LOD of 2.39 nM.[76] Mollaei *et al.* have reported the synthesis of an *n*-dodecylpyridinium chloride-stabilized carbon paste electrode for the detection of folic acid with an LOD of 2.9 nM.[77]

Capelari and co-workers have synthesized a cetylpyridinium bromide-stabilized glassy carbon electrode and used it as a sensor for *o*-nitrophenol and *p*-nitrophenol with LODs of 1.14 μM and 0.59 μM, respectively.[78] Manjunatha *et al.* have investigated the synthesis of a sodium dodecyl sulfate/graphene paste electrode for catechol with an LOD of 0.10 μM.[79] In another report, asodium dodecyl sulfate-stabilized nano-cobalt(II,III) oxide/carbon paste electrode with an LOD of 0.01 μM was prepared by Fekry and co-workers and used for the detection of caffeine.[80]

Mollaei *et al.* have reported the synthesis of an *n*-dodecylpyridinium chloride-stabilized carbon paste electrode for the detection of folic acid and calcium folinate with respective LODs of 7.5 nM and 24.0 nM.[81] Nunes Angelis and co-workers have synthesized a cetylpyridinium bromide-stabilized carbon black paste electrode and used it as a sensor for *tert*-butylhydroquinone and butylated hydroxyanisole with LODs of 0.07 μM and 0.08 μM, respectively[82] (see Table 11.2).

11.4 Limitations of Surfactant Sensors

Some of the surfactants are extremely dangerous to the environment, society, and ecosystems. Some surfactants alter the structure of the root cell membrane and cause harm. As a result, this hampered the transfer of nutrients and water.[83] Ecosystems are impacted by the physiologic, metabolic, pathogenic, and other effects of anionic surfactants such as linear alkylbenzene sulfonates.[84]

Surfactants impair human endocrine functions and have long-term metabolic impacts by reacting with proteins that are already present in the blood and the liver.[85] Skin irritation and respiratory issues are brought on by

Table 11.2 Applications of surfactant sensors.

	Working electrode	Surfactant	Analyte	LOD
1	Hanging mercury drop electrode	Cetyl trimethyl ammonium bromide	Genistein	34.3 nM
2	Hanging mercury drop electrode	Tween 20	Entacapone	0.42 nM
3	Multi-wall carbon nanotube paste electrode	Triton X-100	Aspirin Caffeine	0.084 μM 0.088 μM
4	Glassy carbon electrode	Cetyl trimethyl ammonium bromide	5-Fluorouracil	20.13 nM
5	Boron-doped diamond	Sodium dodecyl sulfate	BaP	2.86 nM
6	Hanging mercury drop electrode	Tween-20	Drotaverine hydrochloride	4.14 nM
7	Cetyl trimethyl ammonium bromide/ etrabromobisphenol	Cetyl trimethyl ammonium bromide	Ascorbic acid Rutin	1.0 μM 4.0 nM
8	Hanging mercury drop electrode	Cetyl trimethyl ammonium bromide	Quinine	0.40 μM
9	Carbon nanotube paste electrode	Cetylpyridinium bromide	Acetaminophen Ascorbic acid	7.1 μM 2.1 μM
10	Grapheme nanosheet paste	Cetyl trimethyl ammonium bromide	Dopamine	0.6 μM
11	Cetyl trimethyl ammonium bromide/carbon paste electrode	Cetyl trimethyl ammonium bromide	Nonylpenol	0.01 μM
12	Dodecyl benzene sulfate/carbon paste electrode	Dodecyl benzene sulfate	Dopamine	0.01 μM
13	Glassy carbon electrode	Cetyl trimethyl ammonium bromide	Testosterone	1.18 nM
14	Cetylpyridinium bromide/single-wall carbon nanotube-COOH/glassy carbon electrode	Cetyl trimethyl ammonium bromide	Morin	28.9 nM

Table 11.2 (*Continued*)

	Working electrode	Surfactant	Analyte	LOD
15	Hanging mercury drop electrode	Cetyl trimethyl ammonium bromide	Cefitizoxime	1.98 μM
16	Multi-wall carbon nanotube-COOH/ sodium dodecyl sulfate/ screenprinted carbon electrode	Sodium dodecyl sulfate	Cisplatin	4.6 μM
17	Multi-wall carbon nanotube cetyl trimethyl ammonium bromide/glassy carbon electrode	Cetyl trimethyl ammonium bromide	4-Amino antipyrine	0.16 nM
18	Ce/carbon paste electrode	Sodium dodecyl sulfate	Nicotine	94.3 nM
19	ZnO/carbon paste electrode	Cetyl trimethyl ammonium bromide	Nilutamide	3.21 nM
20	Glassy carbon electrode	Cetyl trimethyl ammonium bromide	Thiosalicylic acid	0.11 μM
21	Didodecyldimethyl ammonium bromide/expanded graphite paste electrode	Didodecyldimethyl ammonium bromide	Bisphenol	7.1 nM
22	Pyrolytic graphite electrode	Sodium dodecyl sulfate	Rabeprazole sodium	0.18 μM
23	Carbon black/glassy carbon electrode	Cetyl trimethyl ammonium bromide	Mesotrione	0.02 μM
24	Sodium dodecyl sulfate/carbon paste electrode	Sodium dodecyl sulfate	Ethylvanillin Methylvanillin	0.02 μM 0.02 μM
25	Boron-doped diamond	Sodium dodecyl sulfate	Enrofloxacin	0.01 μM
26	Ag-benzalkonium chloride/glassy carbon electrode	Benzalkonium chloride	Tenofovir	2.39 nM
27	Carbon paste electrode	*n*-Dodecylpyridinium chloride	Folic acid	2.9 nM

Table 11.2 (*Continued*)

	Working electrode	Surfactant	Analyte	LOD
28	Glassy carbon electrode	Cetylpyridinium bromide	*o*-Nitrophenol *p*-Nitrophenol	1.14 μM 0.59 μM
29	Sodium dodecyl sulfate/graphene paste electrode	Sodium dodecyl sulfate	Catechol	0.10 μM
30	Nano-cobalt(II,III) oxide/carbon paste electrode	Sodium dodecyl sulfate	Caffeine	0.01 μM
31	Carbon paste electrode	*n*-Dodecylpyridinium chloride	Folic acid Calcium folinate	7.5 nM 24.0 nM
32	Carbon black paste electrode	Cetylpyridinium bromide	*tert*-Butylhydroquinone Butylated hydroxyanisole	0.07 μM 0.08 μM

linear alkylbenzene sulfonates.[86] Surfactants decrease the ability of aquatic biota to withstand development, environmental stress, and reproduction.[87]

Mammals and amphibians are affected by nonylphenol ethoxylates in an estrogenic manner.[88] Surfactants make pollutants more soluble, which promotes eutrophication.[89] According to reports, certain surfactants might irritate human skin and cause ocular and respiratory issues.[90]

11.5 Challenges of Surfactant Sensors

Surfactant sensor technology has advanced by leaps and bounds, thanks to several fascinating nanosensors. Graphene, quantum dots, and nanocomposites are three examples. The first main applications of advanced nanomaterials was for sensing purposes. However, in the post-hype of the field, the commercialization of surfactant sensors as little more than a product must have posed significant obstacles. The most severe problems that must be resolved before these products become commercially viable are growing public and government concerns about the lack of practice standards for evaluating the safety of such detectors with industrial/healthcare needs. The business model for intelligent nanosensors has already grown dramatically as a result of breakthroughs throughout modern tech developments incorporating cyber-physical systems and artificial intelligence. In general, the field of nanosensors is multi-disciplinary, involving science, electronics, and mechanical challenges. To meet the demands of different fields such as pharmaceuticals, therapeutic agents, disease detection, and in cultured cell biomedical diagnostics, the development of next-generation solutions require extensive improvements in selectivity and sensitivity.

11.6 Conclusion

Nano surfactant sensor research aims to create innovative technologies that can significantly contribute to the detection and monitoring of biomedical, biochemical, environmental, agricultural, and food industries. Furthermore, the primary goal is to benefit humanity and society. Because of the advancement of nanotechnology, new horizons for surfactant sensors with dimensions suitable for intracellular use that are submicron-sized, depending on the application, have opened up. Investigation of various special effects, such as dimension, quantum size, and surface effects, should be prioritized. To develop quite advanced as well as affordable surfactant sensors, well-planned multi-disciplinary research involving engineers, bioscience researchers, scientists, and doctors should always be conducted. The future scope for surfactant sensors must be significant, with a slogan of automation, integration, and technological advancement. By incorporating and implementing advanced technological features, this same device could be commercially produced.

References

1. F. Zhang, *et al.*, Adsorption of different types of surfactants on graphene oxide, *J. Mol. Liq.*, 2019, **276**, 338–346.
2. M. G. Chernysheva, *et al.*, Cationic surfactant coating nanodiamonds: Adsorption and peculiarities, *Colloids Surf., A*, 2019, **565**, 25–29.
3. Z. Chang, X. Chen and Y. Peng, The adsorption behavior of surfactants on mineral surfaces in the presence of electrolytes – A critical review, *Miner. Eng.*, 2018, **121**, 66–76.
4. A. Casandra, *et al.*, Adsorption kinetics of the partially dissociated ionic surfactants: The effect of degree of dissociation, *J. Taiwan Inst. Chem. Eng.*, 2018, **92**, 2–7.
5. M. J. Rosen, *Surfactants and Interfacial Phenomena*, Wiley, 2004.
6. M. A. Hegazy, M. Abdallah and H. Ahmed, Novel cationic gemini surfactants as corrosion inhibitors for carbon steel pipelines, *Corros. Sci.*, 2010, **52**(9), 2897–2904.
7. L.-G. Qiu, *et al.*, Synergistic effect between cationic gemini surfactant and chloride ion for the corrosion inhibition of steel in sulphuric acid, *Corros. Sci.*, 2008, **50**(2), 576–582.
8. H.-Y. Wu, *et al.*, Development and validation of an analytical procedure for quantitation of surfactants in dishwashing detergents using ultra-performance liquid chromatography-mass spectrometry, *Talanta*, 2019, **194**, 778–785.
9. S. Yada, *et al.*, Emulsification, solubilization, and detergency behaviors of homogeneous polyoxypropylene-polyoxyethylene alkyl ether type nonionic surfactants, *Colloids Surf., A*, 2019, **564**, 51–58.
10. A. M. Atta, *et al.*, Demulsification of heavy crude oil using new nonionic cardanol surfactants, *J. Mol. Liq.*, 2018, **252**, 311–320.

11. A. M. Abdulraheim, Green polymeric surface active agents for crude oil demulsification, *J. Mol. Liq.*, 2018, **271**, 329–341.
12. C. Xu, *et al.*, Experimental investigation of coal dust wetting ability of anionic surfactants with different structures, *Process Saf. Environ. Prot.*, 2019, **121**, 69–76.
13. L. Alexandrova, *et al.*, Effects of pH on wetting behavior of 'star-like' block copolymer surfactant solutions, *Colloids Surf., A*, 2017, **519**, 78–86.
14. Y. Wang, *et al.*, Halide-free PN ionic liquids surfactants as additives for enhancing tribological performance of water-based liquid, *Tribol. Int.*, 2018, **128**, 190–196.
15. M. Zhao, *et al.*, Study on the synergy between silica nanoparticles and surfactants for enhanced oil recovery during spontaneous imbibition, *J. Mol. Liq.*, 2018, **261**, 373–378.
16. A. M. Howe, *et al.*, Visualising surfactant enhanced oil recovery, *Colloids Surf., A*, 2015, **480**, 449–461.
17. E. Nourafkan, Z. Hu and D. Wen, Controlled delivery and release of surfactant for enhanced oil recovery by nanodroplets, *Fuel*, 2018, **218**, 396–405.
18. L. Liu and J. Song, Voltammetric determination of mefenamic acid at lanthanum hydroxide nanowires modified carbon paste electrodes, *Anal. Biochem.*, 2006, **354**, 22–27.
19. L. Fernández-Llano, M. C. Blanco-Lopez, M. J. Lobo-Castañón, A. J. Miranda-Ordieres and P. Tuñón-Blanco, Determination of diclofenac in urine samples by molecularly-imprinted solid-phase extraction and adsorptive differential pulsevoltammetry, *Electroanalysis*, 2007, **19**, 1555–1561.
20. M. Hajjizadeh, A. Jabbari, H. Heli, A. A. Moosavi-Movahedi and S. Haghgoo, Electrocatalytic oxidation of some anti-inflammatory drugs on a nickel hydroxide-modified nickel electrode, *Electrochim. Acta*, 2007, **53**, 1766–1774.
21. L. Barros, S. Falcão, P. Baptista, C. Freire, M. Vilas-Boas and I. C. F. R. Ferreira, Antioxidant activity of Agaricus sp. mushrooms by chemical, biochemical and electrochemical assays, *Food Chem.*, 2008, **111**, 61–66.
22. A. Masek, M. Zaborski and E. Chrzescijanska, Electrooxidation of flavonoids at platinum electrode studied by cyclic voltammetry, *Food Chem.*, 2011, **127**, 699–704.
23. M. Tubaishat and S. Madria, Sensor networks: an overview, *IEEE Potentials*, 2003, **22**(2), 20–23.
24. J. S. Wilson, *Sensor technology handbook*, Elsevier, 2004.
25. N. Yamazoe, Toward innovations of gas sensor technology, *Sens. Actuators, B*, 2005, **108**(1–2), 2–14.
26. H. Nazemi, A. Joseph, J. Park and A. Emadi, Advanced micro-and nanogas sensor technology: A review, *Sensors*, 2019, **19**(6), 1285.
27. M. Najafzadeh, F. Saberi-Movahed and S. Sarkamaryan, NF-GMDH-Based self-organized systems to predict bridge pier scour depth under debris flow effects, *Mar. Georesour. Geotechnol.*, 2018, **36**(5), 589–602.

28. B. Kuppan and U. Maitra, Instant room temperature synthesis of self-assembled emission-tunable gold nanoclusters: million-fold emission enhancement and fluorimetric detection of Zn^{2+}, *Nanoscale*, 2017, **9**(40), 15494–15504.
29. N. Barsan, D. Koziej and U. Weimar, Metal oxide-based gas sensor research: How to?, *Sens. Actuators, B*, 2007, **121**(1), 18–35.
30. J. Singh, N. Batra and R. C. Sobti, *J. Ind. Microbiol. Biotechnol.*, 2004, **31**, 51–58.
31. L. Fu, V. P. Dravid and D. L. Johnson, *Appl. Surf. Sci.*, 2001, **181**, 173–178.
32. A. Bodour, C. Guerrero-Barajas, B. V. Jiorle, M. E. Malcomson, A. K. Paull, A. Somogyi, L. N. Trinh, R. B. Bates and R. M. Maier, *Appl. Environ. Microbiol.*, 2004, **70**, 114–120.
33. J. Najeeb, U. Farwa, F. Ishaque, H. Munir, A. Rahdar, M. F. Nazar and M. N. Zafar, Surfactant stabilized gold nanomaterials for environmental sensing applications–A review, *Environ. Res.*, 2022, **208**, 112644.
34. R. Jain, A. Haque and A. Verma, Voltammetric quantification of surfactant stabilized curcumin at MWCNT/GCE sensor, *J. Mol. Liq.*, 2017, **230**, 600–607.
35. S. Kaushal, N. Priyadarshi, A. K. Pinnaka, S. Soni, A. Deep and N. K. Singhal, Glycoconjugates coated gold nanorods based novel biosensor for optical detection and photothermal ablation of food borne bacteria, *Sens. Actuators, B*, 2019, **289**, 207–215.
36. A. Kapur, F. Aldeek, X. Ji, M. Safi, W. Wang, A. Del Cid, O. Steinbock and H. Mattoussi, Self-assembled gold nanoparticle–fluorescent protein conjugates as platforms for sensing thiolate compounds via modulation of energy transfer quenching, *Bioconjugate Chem.*, 2017, **28**(2), 678–687.
37. S. Liu, J. Li and X. Mao, Stripping voltammetric determination of zirconium with complexing preconcentration of zirconium(IV) at a morin-modified carbon paste electrode, *Electroanalysis*, 2003, **15**, 1751–1755.
38. Q. Lu, W. Zhang, Z. Wang, G. Yu, Y. Yuan and Y. Zhou, A facile electrochemical sensor for nonylphenol determination based on the enhancement effect of cetyltrimethylammonium bromide, *Sensors*, 2013, **13**, 758–768.
39. I. Martynov, Y. Kuzishchin, D. Dovzhenko, G. Kotkovskii and A. Chistyakov, Ionization of the nitroaromatic compounds in an ion mobility spectrometer with an ion source based on porous silicon under laser irradiation, *Phys. Procedia*, 2015, **73**, 163–167.
40. F. J. Ibañez and F. P. Zamborini, Chemiresistive Sensing of Volatile Organic Compounds with Films of Surfactant-Stabilized Gold and GoldSilver Alloy Nanoparticles, *ACS Nano*, **8**(2), 1543–1552.
41. C. Raril and J. G. Manjunatha, Voltammetric determination of anthrone using cetyl trimethyl ammonium bromide surfactant modified carbon paste electrode, *Biomed. J. Sci. Tech. Res.*, 2018, **11**, 8560–8564.
42. P. Y. Khashaba, H. R. H. Ali and M. M. El-Wekil, Square wave stripping voltammetric determination of pantoprazole in rabbit plasma using surfactant-based pencil graphite electrode, *J. Anal. Chem.*, 2019, **74**, 609–616.

43. J. G. Manjunatha, A surfactant enhanced graphene paste electrode as an effective electrochemical sensor for the sensitive and simultaneous determination of catechol and resorcinol, *Chem. Data Collect.*, 2020, **25**, 100331.
44. P. Deng, Z. Xu and Y. Feng, Highly sensitive and simultaneous determination of ascorbic acid and rutin at an acetylene black paste electrode coated with cetyltrimethyl ammonium bromide film, *J. Electroanal. Chem.*, 2012, **683**, 47–54.
45. S. R. Sataraddi and S. T. Nandibewoor, Voltammetric-oxidation and determination of 5-flurouracil and its analysis in pharmaceuticals and biological fluids at glassy carbon electrode mediated by surfactant cetyltrimethyl ammonium bromide, *Der Pharma Chem.*, 2011, **3**, 253.
46. D. S. Nayak and N. P. Shetti, Voltammetric response and determination of an anti-inflammatory drug at a cationic surfactant-modified glassy carbon electrode, *J. Surfactants Deterg.*, 2016, **19**, 1071–1079.
47. P. B. Deroco, B. C. Lourencao and O. Fatibello-Filho, The use of modified electrode with carbon black as sensor to the electrochemical studies and voltammetric determination of pesticide mesotrione, *Microchem. J.*, 2017, **133**, 188–194.
48. H. Zhai, S. Wang, J. Zhou, J. Pan, Y. Tong and Q. Mei, *et al.*, A simple and sensitive electrochemical sensor for 3-nitrotyrosine based on electrochemically anodic pretreated glassy carbon electrode in anionic surfactant medium, *J. Electrochem. Soc.*, 2019, **166**, B1426–B1433.
49. P. Y. Khashaba, H. R. Ali and M. El-Wekil, Surfactant based voltammetric analysis of anti-ulcer drug in real samples, *Anal. Baioanal. Electrochem.*, 2017, **9**, 729–740.
50. J. A. Rather and R. Jain, Stripping voltammetric detection of nephrotoxic drug cefitizoxime in wastewater, *Anal. Chem. Res.*, 2015, **4**, 13–19.
51. D. K. Fogliatto, A. M. J. Barbosa and V. S. Ferreira, Voltammetric determination of the phytoestrogen genistein in soy flours and soy based supplements using cationic surfactant cetyltrimetylammonium bromide, *Colloids Surf., B*, 2010, **78**, 243–249.
52. R. Jain, R. K. Yadav and A. Dwivedi, Square-wave adsorptive stripping voltammetric behaviour of entacapone at HMDE and its determination in the presence of surfactants, *Colloids Surf., A*, 2010, **359**, 25–30.
53. B. J. Sanghavi and A. K. Srivastava, Simultaneous voltammetric determination of acetaminophen, aspirin and caffeine using an in situ surfactant-modified multiwalled carbon nanotube paste electrode, *Electrochim. Acta*, 2010, **55**, 8638–8648.
54. S. R. Sataraddi and S. T. Nandibewoor, Voltammetric-oxidation and determination of 5-flurouracil and its analysis in pharmaceuticals and biological fluids at glassy carbon electrode mediated by surfactant cetyltrimethyl ammonium bromide, *Der Pharma Chem.*, 2011, **3**, 253.
55. Y. Yardım, A. Levent, E. Keskin and Z. Şentürk, Voltammetric behavior of benzo [a] pyrene at boron-doped diamond electrode: a study of its determination by adsorptive transfer stripping voltammetry based on

the enhancement effect of anionic surfactant, sodium dodecylsulfate, *Talanta*, 2011, **85**, 441–448.

56. R. Jain and J. A. Rather, Voltammetric behaviour of drotaverine hydrochloride in surfactant media and its enhancement determination in Tween-20, *Sens. Actuators, B*, 2011, **82**, 333–339.
57. P. Deng, Z. Xu and Y. Feng, Highly sensitive and simultaneous determination of ascorbic acid and rutin at an acetylene black paste electrode coated with cetyltrimethyl ammonium bromide film, *J. Electroanal. Chem.*, 2012, **683**, 47–54.
58. R. A. Dar, P. K. Brahman, S. Tiwari and K. S. Pitre, Electrochemical studies of quinine in surfactant media using hanging mercury drop electrode: a cyclic voltammetric study, *Colloids Surf., B*, 2012, **98**, 72–79.
59. E. H. Duarte, L. T. Kubota and C. R. T. Tarley, Carbon nanotube based sensor for simultaneous determination of acetaminophen and ascorbic acid exploiting multiple response optimization and measures in the presence of surfactant, *Electroanalysis*, 2012, **24**, 2291–2301.
60. S. Q. Liu, W. H. Sun and F. T. Hu, Graphene nano sheet-fabricated electrochemical sensor for the determination of dopamine in the presence of ascorbic acid using cetyltrimethylammonium bromide as the discriminating agent, *Sens. Actuators, B*, 2012, **173**, 497–504.
61. Q. Lu, W. Zhang, Z. Wang, G. Yu, Y. Yuan and Y. Zhou, A facile electrochemical sensor for nonylphenol determination based on the enhancement effect of cetyltrimethylammonium bromide, *Sensors*, 2013, **13**, 758–768.
62. S. S. Shankar, B. K. Swamy, B. N. Chandrashekar and K. J. Gururaj, Sodium do-decyl benzene sulfate modified carbon paste electrode as an electrochemical sensor for the simultaneous analysis of dopamine, ascorbic acid and uric acid: a voltammetric study, *J. Mol. Liq.*, 2013, **177**, 32–39.
63. A. Levent, A. Altun, Y. Yardım and Z. Şentürk, Sensitive voltammetric determination of testosterone in pharmaceuticals and human urine using a glassy carbon electrode in the presence of cationic surfactant, *Electrochim. Acta*, 2014, **128**, 54–60.
64. G. Ziyatdinova, E. Ziganshina and H. Budnikov, Electrooxidation of morin on glassy carbon electrode modified by carboxylated singlewalled carbon nanotubes and surfactants, *Electrochim. Acta*, 2014, **145**, 209–216.
65. J. A. Rather and R. Jain, Stripping voltammetric detection of nephrotoxic drug cefitizoxime in wastewater, *Anal. Chem. Res.*, 2015, **4**, 13–19.
66. E. M. Materon, A. Wong, S. I. Klein, J. Liu and M. D. Sotomayor, Multiwalled carbon nanotubes modified screen-printed electrodes for cisplatin detection, *Electrochim. Acta*, 2015, **158**, 271–276.
67. J. I. Gowda, A. T. Buddanavar and S. T. Nandibewoor, Fabrication of multiwalled carbon nanotube-surfactant modified sensor for the direct determination of toxic drug 4-aminoantipyrine, *J. Pharm. Anal.*, 2015, **5**, 231–238.
68. A. M. Fekry, S. M. Azab, M. Shehata and M. A. Ameer, A novel electrochemical nicotine sensor based on cerium nanoparticles with anionic surfactant, *RSC Adv.*, 2015, **5**, 51662–51671.

69. Y. Temerk, H. Ibrahim and N. Farhan, Square wave adsorptive stripping voltammetric determination of anticancer drug nilutamide in biological fluids using cationic surfactant cetyltrimethylammonium bromide, *Anal. Methods*, 2015, **7**, 9137–9144.
70. D. S. Nayak and N. P. Shetti, Voltammetric response and determination of an anti-inflammatory drug at a cationic surfactant-modified glassy carbon electrode, *J. Surfactants Deterg.*, 2016, **19**, 1071–1079.
71. J. Zhang, S. Ma, W. Wang and Z. Chen, Electrochemical sensing of bisphenol A by a didodecyldimethylammonium bromidemodified expanded graphite paste electrode, *J. AOAC Int.*, 2016, **99**, 1066–1072.
72. P. Y. Khashaba, H. R. Ali and M. El-Wekil, Surfactant based voltammetric analysis of anti-ulcer drug in real samples, *Anal. Baioanal. Electrochem.*, 2017, **9**, 729–740.
73. P. B. Deroco, B. C. Lourencao and O. Fatibello-Filho, The use of modified electrode with carbon black as sensor to the electrochemical studies and voltammetric determination of pesticide mesotrione, *Microchem. J.*, 2017, **133**, 188–194.
74. A. S. Farag, M. Sýs, T. Hájek and K. Vytřas, Voltammetric determination of ethylvanillin and methylvanillin sum at carbon paste electrode modified by sodium dodecyl sulfate in selected foodstuffs, *Monatsh. Chem.*, 2018, **149**, 1945–1953.
75. F. Dönmez, Y. Yardım and Z. Şentürk, Electroanalytical determination of enrofloxacin based on the enhancement effect of the anionic surfactant at anodically pretreated boron-doped diamond electrode, *Diamond Relat. Mater.*, 2018, **84**, 95–102.
76. G. Ozcelikay, B. Dogan-Topal and S. A. Ozkan, An electrochemical sensor based on silver nanoparticles-benzalkonium chloride for the voltammetric determination of antiviral drug tenofovir, *Electroanalysis*, 2018, **30**, 943–954.
77. M. Mollaei, S. M. Ghoreishi and A. Khoobi, Electrochemical investigation of a novel surfactant for sensitive detection of folic acid in pharmaceutical and biological samples by multivariate optimization, *Measurement*, 2019, **145**, 300–310.
78. T. B. Capelari, A. C. Pereira, L. L. Gonçalves de Oliveira and C. R. Teixeira Tarley, Sensitive simultaneous determination of onitrophenol and *p*-nitrophenol in water by surfactant-mediated differential pulse voltammetry, *Anal. Lett.*, 2019, **52**, 1462–1476.
79. J. G. Manjunatha, A surfactant enhanced graphene paste electrode as an effective electrochemical sensor for the sensitive and simultaneous determination of catechol and resorcinol, *Chem. Data Collect.*, 2020, **25**, 100331.
80. A. M. Fekry, M. Shehata, S. M. Azab and A. Walcarius, Voltammetric detection of caffeine in pharmacological and beverages samples based on simple nano-Co(ii, iii) oxide modified carbon paste electrode inaqueous and micellar media, *Sens. Actuators, B*, 2020, **302**, 127172.

81. M. Mollaei, S. M. Ghoreishi and A. Khoobi, Multivariate optimization and validation of a new procedure for simultaneous determination of folic acid and folinic acid based on enhancement effect of n-dodecylpyridinium chloride, *Microchem. J.*, 2020, **154**, 104653.
82. P. Nunes Angelis, J. de Cássia Mendonça, L. Rianne da Rocha, T. Boareto Capelari, M. Carolyne Prete and M. Gava Segatelli, Feasibility of a nano-carbon black paste electrode for simultaneous voltammetric determination of antioxidants in food samples and biodiesel in the presence of surfactant, *Electroanalysis*, 2020, **32**, 1198–1207.
83. K. Jardak, P. Drogui and R. Daghrir, Surfactants in aquatic and terrestrial environment: occurrence, behavior, and treatment processes, *Environ. Sci. Pollut. Res.*, 2016, **23**, 3195–3216.
84. B. Petrie, R. Barden and B. Kasprzyk-Hordern, A review on emerging contaminants in wastewaters and the environment: current knowledge, understudied areas and recommendations for future monitoring, *Water Res.*, 2015, **72**, 3–27.
85. C. C. Borghi, M. Fabbri, M. Fiorini, M. Mancini and P. L. Ribani, Magnetic removal of surfactants from wastewater using micrometric iron oxide powders, *Sep. Purif. Technol.*, 2011, **83**, 180–188.
86. M. C. Collivignarelli, M. Carnevale Miino, M. Baldi, S. Manzi, A. Abbà and G. Bertanza, Removal of non-ionic and anionic surfactants from real laundry wastewater by means of a full-scale treatment system, *Process Saf. Environ. Prot.*, 2019, **132**, 105–115.
87. M. Hampel, A. Mauffret, K. Pazdro and J. B. Andalusian, Anionic surfactant linear alkylbenzene sulfonates (LAS) in sediments from the Gulf of Gdansk (southern Baltic Sea, Poland) and its environmental implications, *Environ. Monit. Assess.*, 2012, **184**, 6013–6023.
88. C. C. Borghi, M. Fabbri, M. Fiorini, M. Mancini and P. L. Ribani, Magnetic removal of surfactants from wastewater using micrometric iron oxide bpowders, *Sep. Purif. Technol.*, 2011, **83**, 180–188.
89. A. Zanoletti, S. Federici, L. Borgese, P. Bergese, M. Ferroni, L. E. Depero and E. Bontempi, Embodied energy as key parameter for sustainable materials selection: the case of reusing coal fly ash for removing anionic surfactants, *J. Cleaner Prod.*, 2017, **141**, 230–236.
90. G. G. Ying, Fate, behavior and effects of surfactants and their degradation products in the environment, *Environ. Int.*, 2006, **32**, 417–431.

Subject Index